Wolcott Cronk Foster

A Treatise on Wooden Trestle Bridges

According to the Present Practice on American Railroads

Wolcott Cronk Foster

A Treatise on Wooden Trestle Bridges
According to the Present Practice on American Railroads

ISBN/EAN: 9783744666770

Printed in Europe, USA, Canada, Australia, Japan

Cover: Foto ©berggeist007 / pixelio.de

More available books at **www.hansebooks.com**

A TREATISE

ON

WOODEN TRESTLE BRIDGES

ACCORDING TO THE PRESENT PRACTICE
ON AMERICAN RAILROADS.

BY

WOLCOTT C. FOSTER.

NEW YORK:

JOHN WILEY & SONS,

53 East Tenth Street.

1891.

Press,
Printers
88 Pearl Street,
New York.

PREFACE.

In collecting the data for this work, a circular letter was sent to each chief engineer throughout the country of whom the author could obtain the address. These letters met with many hearty responses, and resulted in the collection of a very complete set of plans of the standard trestles in use on the different roads.

Tables embodying the details of all the different parts were compiled, and the deductions made from these. Every effort has been put forth to make the work as valuable and complete as possible, without making it too bulky. Neither time, pains, nor expense have been spared in its preparation.

As far as possible credit has been given in the body of the work to the originators of any special design; but as oversights may have unintentionally occurred, a list of those engineers who have aided the author is appended.

It is the earnest hope of the author that the results of his labors will prove worthy of the courtesy and aid so generously extended to him by the members of the profession at large, many of whom were perfect strangers.

LIST OF THE ENGINEERS TO WHOM THE AUTHOR IS INDEBTED FOR AID IN PREPARING THIS WORK.

Alger, Chas. E.	Fitch, A. B.	Nelson, J. P.
Ansart, Felix.	Fratt, F. W.	Nettleton, G. A.
Bates, Onward.	Gore, Th.	Nicholson, G. B.
Becker, M. J.	Greenleaf, J. L.	Patton, E. B.
Berg, Walter G.	Griggs, J.	Perris, Fred. T.
Bissel, F. E.	Hawks, J. D.	Reed, A. L.
Blunt, Jno. E.	Howe, W. B. W., Jr.	Rich, W. W.
Boiton, C. M.	Hoyt, Wm. E.	Riffle, F.
Booker, B. F.	Kennedy, H. A.	Rowe, S. M.
Bowen, A. L.	Kriegshaber, V. H.	Sage, I. Y.
Briggs, R. E.	Levings, Chas.	Schenck, A. A.
Buxton, C.	Lum, D. W.	Smith, P. A.
Canfield, E.	Martin, M. A.	Spofford, Parker.
De Caradene, A.	McVean, J. J.	Swift, A. J.
Curtis, F. S.	Miller, N. D.	Weeks, I. S. P.
Davery, R. A.	Mills, A. L.	Wheeler, D. M.
Dick, H. B.	Molesworth, A. N.	White, H. F.
Dorsey, W. H., Jr.	Monroe, J. A.	Whittemore, D. J.
Elliott, R. H.	Montfort, R.	Woods, J. E.
Fisher, J. B.	Morton, T. L.	Zook F. K.

CONTENTS.

PART I.

CHAPTER I.

INTRODUCTION.

CHAPTER II.

PILE-BENTS.

HAPTER III.

PILE-DRIVERS.

CHAPTER IV.

FRAMED BENTS.

CHAPTER V.

FLOOR SYSTEM.

CHAPTER VI.

BRACING, COMPOUND-TIMBER TRESTLES, HIGH TRESTLES, TRESTLES ON CURVES, AND MISCELLANEOUS TRESTLES.

CHAPTER VII.

IRON DETAILS.

CHAPTER VIII.

CONNECTION WITH THE EMBANKMENT AND PROTECTION AGAINST ACCIDENTS.

CHAPTER IX.

FIELD-ENGINEERING AND ERECTION.

CHAPTER X.

PRESERVATION AND STANDARD SPECIFICATIONS.

CHAPTER XI.

BILLS OF MATERIAL, RECORDS AND MAINTENANCE.

PART II.

SECTION I.

PILE-TRESTLES.

SECTION II.

FRAMED TRESTLES.

TECHNICAL TERMS AND NAMES.

THE following list gives the names and their synonyms of some of the more important parts of wooden trestles. In connection with this list see Figs. 1 and 2, to which the numbers opposite the names refer.

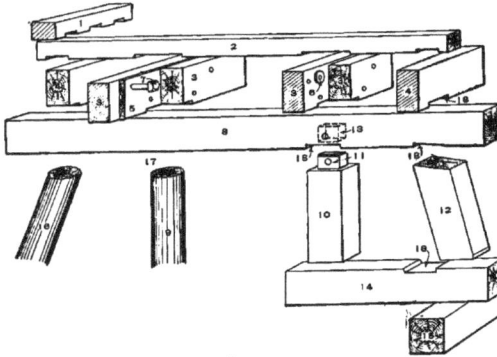

FIG. 1.

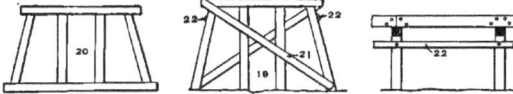

FIG. 2.

Bent, Framed, 20. (See page 24.)
    Pile, 19.   "   "   6
    Cluster.   "   "   41
Bent Brace, see Sway-brace.
Block, see Sub-sill.
Bolster, see Corbel.
Cap, 3. (See page 12.)
Chord, see Stringer.
Corbel, Bolster. (See page 31.)
Cross-tie, 2.   "   "   35
Cut-off, 17.   "   "   11
Dapping, see Notching.
Fender, Guard-rail, 1. (See page 35.)
Gaining, see Notching.
Girt, see Longitudinal Brace.
Girder, see Stringer.
Guard-rail, Fender, Ribbands, 1. (See page 35.)
Jack-stringer, see Stringer.
Longitudinal Brace, Girt, Waling-strip, 22. (See
Mortise, 13.   [page. 39)
Mud-sill, see Sub-sill.
Notching, Gaining, Dapping, 18. (See page 30.)

Outside Stringer, see Stringer.
Packing-block, Packing piece, 5. (See page 32.)
Packing-bolt, 7.   "   "   48
Packing-piece, see Packing-block.
Packing-washers, see Separator.
Piles, Batter, Inclined Brace, 16. (See page 7.)
    Vertical, Plumb, Upright, 9.   "   "   7
Posts, Batter, Inclined, 12.   "   "   28
    Vertical, Plumb, Upright, 10   "   "   28
Ribbands, see Guard-rail.   [pages 32 and 50.)
Separator, Packing-washer, Thimble Spool, 6. (See
Sill, 14. (See page 27.)
Spool, see Separator.
Stringer, Chord, Girder.
    Track, 3.   (See page 32.)
    Outside, Jack, 4. "   "   34
Sub-sill, Mud-sill, Blocks, 15. (See page 25.)
Sway-brace, Bent Brace, 21.   "   "   39
Tenon, 11.   "   "   12
Thimble, see Separator.
Track-stringer, see Stringer.
Waling-strip, see Longitudinal Brace.

ABBREVIATIONS.

A. & P. R. R.; Atlantic & Pacific Railroad.

B., C. R. & N. R. R.; Burlington, Cedar Rapids & Northern Railroad.

B. & M. R. R. R. in Neb.; Burlington & Missouri River Railroad in Nebraska.

C. & A. Ry.; Chicago & Atlantic Railway.

C., B. & Q. R. R.; Chicago, Burlington & Quincy Railroad.

C., C. & C. R. R.; Charleston, Cincinnati & Chicago Railroad.

C., M. & St. P. Ry.; Chicago, Milwaukee & St. Paul Railway.

C., N. O. & T. P. Ry.; Cincinnati, New Orleans & Texas Pacific Railway.

C. & S. Ry.; Charleston & Savannah Railway.

C. & W. M. Ry.; Chicago & West Michigan Railway.

D., T. & Ft. W. R. R.; Denver, Texas & Fort Worth Railroad.

G., C. & S. F. R. R.; Gulf, Colorado & Santa Fe Railroad.

K C., Ft. S. & M. R. R.; Kansas City, Fort Scott & Memphis Railroad.

K., G. B. & W. R. R.; Kewaunee, Green Bay & Western Railroad.

L. & N. R. R.; Louisville & Nashville Railroad.

M., K. & T. Ry.; Missouri, Kansas & Texas Railway.

N. Y., P. & B. R. R.; New York, Providence & Boston Railroad.

N. Y., W. S. & B. R. R.; New York, West Shore & Buffalo Railroad.

R. & D. R. R.; Richmond & Danville Railroad.

St. P., M. & M. R. R.; St. Paul, Minneapolis & Manitoba Railroad.

S. F. & N. P. R. R.; San Francisco & North Pacific Railroad.

S., F. & W. Ry.; Savannah, Florida & Western Railway.

T., St. L. & K. C. R. R.; Toledo, St. Louis & Kansas City Railroad.

A TREATISE ON WOODEN TRESTLE BRIDGES.

PART I.

CHAPTER I.

INTRODUCTION.

THE amount of Wooden Trestling in this country is very large, but few probably realizing its extent unless they have thoroughly studied the subject. At the present time there are about 2400 miles of single-track railway-trestle in the United States,* of which we can consider about one quarter as only temporary, to be replaced by embankment. " Of the remaining 1800 miles, at least 800 miles will be maintained in wood." This 2400 miles is composed of about 150,000 separate structures having about 730,000 spans or more. Table I gives the general data as to the amount of bridges and trestles, and the average rate per mile of track on some of the more important systems.

TABLE I.

Amount of Bridging and Trestling in Different Parts of the United States, and the Rate per Mile of Track.

(COOPER'S TABLE NO. 3.)

System of Railroad or State.	Miles of Road.	Total Length of Bridges and Trestles in feet.	Lin. ft. of Bridges and Trestles per Mile of Road.
New York Central and West Shore Railroads,	2,894	364,722	126
New York, Lake Erie & Western Railroad,	1,514	95,509	63
Other roads in New York,	3,586	445,900	130
Roads in Pennsylvania,	4,352	336,957	77
" " New England,	2,199	176,700	80
Wabash System,	1,636	160,025	98
Missouri Pacific System,	4,707	566,953	120
Chicago, Milwaukee & St. Paul Railroad,	5,727	614,736	107
St. Louis & San Francisco Railway,	1,441	130,075	90
Denver & Rio Grande Railroad,	1,458	102,195	70
Union Pacific Railroad,	4,754	276,032	58
Louisville & Nashville Railroad,	2,495	322,679	123
Queen and Crescent System,	1,139	299,222	231 †
Roads in Illinois,	8,539	707,535	83
" " Michigan,	4,151	249,345	60
" " Iowa,	7,778	1,049,386	135
Central Railroad and Banking Co. of Georgia,	1,487	173,975	117
Totals,	59,857	6,071,946	101

* In the first part of this chapter a considerable portion of the matter relating to statistics was taken from a paper by Theodore Cooper on American Railroad Bridges, Trans. Amer. Soc. C. E., July 1889.

S. C. Clark in *Scribner's Magazine* for June 1888 gives the length of wooden trestling in the United States at about 2127 miles.

† Includes the crossing of Lake Pontchartrain, a trestle 22 miles long.

"It shows that the relative amount of bridges and trestles varies in different localities from 58 feet per mile to 231 feet per mile. This last, however, is excessive from including the crossing of Lake Pontchartrain, near New Orleans, on a trestle 22 miles long. Omitting this, we would get only 162 feet per mile as the maximum."

"These variations are not entirely due to geographical location, as might appear at first thought. They are also affected by principles governing the original location of each road or division of a system. The alignment and grade may have been sacrificed to the avoidance of bridges and trestles, or the contrary."

"From the large mileage covered by our table, we can rely with considerable confidence upon our average. Taking, therefore, 100 feet per mile as our basis of estimate, we have for the 160,000 miles of railroad in the United States, 16,000,000 feet or 3030 miles of bridges and trestles. Table II gives the distribution of the bridges upon 26,000 miles of railroad into spans of different length."

TABLE II.

Distribution of Bridges and Trestles in Spans of Different Lengths, in Totals of Lineal Feet.

(COOPER'S TABLE No. 4.)

Miles of Road.	Trestles and Spans under 20 feet.	Spans 20 to 50 feet.	Spans 50 to 100 feet.	Spans 100 to 150 feet.	Spans 150 to 200 feet.	Spans 200 to 300 feet.	Spans 300 to 400 feet.	Spans 400 to 500 feet.	Spans over 500 feet.	Total.	Average per Mile of Road.
26,288	2,299,758	85,181	94,165	149,121	80,551	29,542	5,677	1,211	1,040	2,746,246	104.7

"Using this as a basis of estimate, the 3030 miles of trestles and bridges in the United States should be distributed as follows:

	Miles.	No of Spans.
Trestles and Spans under 20 feet, .	2,424	727,200
Spans 20 to 50 feet, ...	121	18,150
" 50 to 100 "	130	9,100
" 100 to 150 feet, . . .	190	8,000
" 150 to 200 " . . .	109	3,300
" over 200 feet . . .	56	1,150
	3,030	766,900

The above includes all bridges of either wood or iron."

In order that we may more fully comprehend the meaning of these figures, let us find the money value. Taking the amount of trestling at an even 2400 miles = 12,672,000 lineal feet. Now about $6 per lineal foot is a fair average for the cost, with timber at $30 per M, B. M., erected. At this rate the trestling represents an expenditure of $76,032,000. With an average life of eight years, which is perhaps a little too long taking everything into consideration, the annual expenditure for repairs and renewals alone amount to $9,504,000, necessitating the use of 316,800,000 feet (B. M.) of timber. Capitalizing this annual expenditure at 4 per cent, we find it represents $237,600,000. Now, if, as Cooper estimates, two thirds of the total amount of trestling is capable of being replaced, we will be justified in spending $168,400,000, with interest at 4 per cent, in accomplishing this end. As one third, or 800 miles, will, of necessity, remain as it is, there will be a continual annual expenditure of $3,168,000 for repairs and renewals, requiring 105,600,000 feet (B. M.) of timber, and representing a capitalized value of $79,200,000 at 4 per cent. These figures do not take into account any increase in the mileage from the building of new roads. From the above we can see what an enormous annual drain there is upon our forests merely for the maintenance of what has been considered one of the smaller and less important of railway properties, and these figures, large as they are, are rather too low than otherwise.

Converting these capitalized values into earthwork, we find that we could build the following number of miles of embankment, twenty feet high, complete, ready for the rails

TABLE II (a).

Amount of Embankment, 20 Feet high, which can be built for the Capitalized Value of the Annual cost of Repairs for the 1600 miles of Replaceable Trestle.

Ties, ...   2640 per mile, @ 45 cents each.
Ballast, ...   2738 cubic yards per mile, @ 50 cents per cubic yard.

Cost of earth per cubic yard, ...	16 c.	18 c.	20 c.	22 c.
Miles, ...	5,262	4,722	4,283	3,918

But of this replaceable two thirds or 1600 miles, only about 600 miles is capable of being replaced by embankment. Now taking the cost of replacing this 600 miles in, say, 20-foot earth embankment, we would have the amount left as indicated in Table III. for replacing the remaining 1000 miles with other permanent structures.

TABLE III.

Showing Cost of 600 Miles of 20-foot Embankment Complete, and Balances.

Capitalized value 4%, ...	$158,400,000.			
Cost of earth per cubic yard, ...	16 c.	18 c.	20 c.	22 c.
Cost of 600 miles of road complete except rails, ...	$18,058,800	$20,124,000	$22,189,800	$24,255,600
Balance, applicable to replacing 1000 miles of trestles by other structures such as iron or masonry bridges, etc., ...	$140,341,200	$138,276,000	$136,210,200	$134,144,400

NOTE.—In this table the same data have been used as in Table II. (a), viz.:

Cost of earth per cubic yard, ...	16 c.	18 c.	20 c.	22 c.
Ties per mile, ...	$1,188	$1,188	$1,188	$1,188
Ballast, 1' x 14', per mile, ...	1,369	1,369	1,369	1,369
Embankment, 14' x 20', (172, 128 cubic yards), per mile,	27,541	30,983	34,426	37,869
Total cost per mile of road, ...	$30,098	$33,540	$36,983	$40,426

There are many other reasons, in addition to the above, which would justify a much larger expenditure than this to secure the replacement of the trestles.

Notwithstanding the great importance of the subject, and the fact that a large part of the expense of building many new roads is chargeable to these structures, no effort of any moment has been made to collect and publish together any considerable amount of data relating to it; the most extensive and important paper so far published on trestling probably being that by Prof. Jameson in *The Engineering and Railroad Journal* for the latter part of 1889 and early part of 1890.

A *good* wooden structure is preferable to the *cheap* iron ones that some roads seem determined to erect. They have proven the salvation of many a new enterprise, when, had it been absolutely necessary to resort to the use of stone or iron, or to make enormous fills, the project must have been abandoned on account of the lack of capital wherewith to erect these costly works. Wooden trestles for the most part are, of course, built with the idea of their being only temporary expedients, to be replaced in time, as rapidly as the finances of the company may permit, by something more permanent. However, a well-built trestle of good material will last a long time, depending to a certain extent on climatic conditions. If properly designed and cared for they form an efficient portion of the roadway. They require constant watching; and the moment any sign of weakness or injurious amount of decay appears it should be remedied immediately. The inspection should be regular and frequent, and placed in careful, trustworthy, and competent hands. It is the practice on some roads, and a very pernicious one which cannot be too strongly condemned, to allow these structures to deteriorate until they are just about ready to fall every time a train passes over them, before the management will attempt to make any repairs, thinking perhaps that they are accomplishing wonders in the way of economy. In consequence of this way of conducting affairs there is scarcely a week that passes but we read of one or more trestle accidents.

The height at which it becomes more economical to replace embankment by trestling varies in different locations, depending upon the cost of lumber, labor, and the facilities for obtaining, and the nature of, the material wherewith to make the fill (see Table IV). There are many places where an embankment would be altogether out of the question, such as across water-ways, swamps with deep, soft mud, etc.; and the only resort then is either to wooden or iron structures.

TABLE IV.

Showing Approximate Relative Cost of Embankment and Trestle in sections of 100 feet, excluding Rails, Ties, and Ballast on former, and Rails, Guard-rails, and Ties on latter.

Height from Surface of Ground to Grade (Sub-grade) in Feet.	Embankment per Cubic Yard in Cents. Road-bed 14 feet wide, Slope 1½ to 1.				TRESTLE. Timber erected (including iron) per M., B. M.					
					Pile-trestle—piling 35 c. per lin. ft. in place; average penetration 10 ft.			Framed Trestles.		
	16	18	20	22	$30	$35	$40	$30	$35	$40
5	$64	$72	$80	$88	$376	$407	$439	$283	$330	$378
10	113	127	141	155	441	476	512	385	449	514
15	325	366	406	447	508	544	580	464	541	618
20	521	587	652	718	576	613	651	541	631	721
25	764	859	955	1050	748	803	858	796	928	1060
30	1049	1180	1312	1443	816	872	928	872	1017	1163
35	1380	1552	1725	1897	990	1065	1140	1058	1234	1410
40	1754	1974	2193	2412	1057	1132	1218	1133	1322	1510
45	2174	2446	2717	2989				1202	1404	1606

If it is necessary to place a masonry structure through a portion of the embankment, then the height at which it will be more economical to build a trestle will be considerably lowered.

While the cost of an embankment increases in a vastly greater ratio than its height, the cost of trestling does not increase nearly as rapidly as its height, especially when under fifty feet. This fact is very clearly shown in Table V.

TABLE V.

Cost of Pile and Framed Trestles complete, including Floor Systems, for Different Heights, in Sections of 100 feet.

Height.	Pile.			Framed.		
	$30	$35	$40	$30	$35	$40
5	$546	$605	$665	$453	$528	$604
10	611	674	738	555	647	740
15	678	742	806	634	739	844
20	746	811	877	711	829	947
25	918	1001	1084	966	1126	1286
30	986	1070	1154	1042	1215	1389
35	1160	1263	1366	1228	1432	1636
40	1227	1332	1444	1303	1520	1736
45				1372	1602	1832

A few engineers have advocated the use of mathematics in the designing of trestles, but as wood is an article whose strength and properties vary rather widely with every piece, no dependence whatever can be placed on the results, and such practice is to be condemned. It is far wiser to merely follow one's judgment and the results of the experience of others as to the proper proportioning of the various parts, gained from experience in dealing with the wood, than to follow any special set of mathematical formulas.

It will probably be impossible to ever thoroughly standardize the plan of trestle design, as there are about as many styles as designers. There also appears to be a tendency to draw up the specifications relating to this subject in a loose and slipshod manner. This is to be much regretted ; as great care and attention in proportion to the importance of the subject should be devoted to this part of the railway's property as to any other.

All structures of this kind, especially those of any extent whatever, should be protected by a re-railing device of some kind, though there are still few that are so protected. Not only should this be the case, but they should also have some kind of fire protection and convenient means for the extinguishment of fires.

There may be said to be two general classes of wooden trestle bridges, namely, those in which the bents consist exclusively of piles and a cap and hence are known as PILE-TRESTLES, and those in which the timbers composing the bents are squared, and framed together, and known as FRAMED TRESTLES. Pile-trestles are seldom used for heights above thirty feet, and it is only occasionally that they are built as high as this. Framed trestles may be of almost any height, though requiring special designs for those above thirty to forty feet. For trestles above forty feet high the cluster-bent form seems to be quite a favorite class of design.

CHAPTER II.

PILE-BENTS.

PILE-BENTS are generally used where the ground is quite soft, and may either occasionally or constantly be covered with water; also where the distance from the rails to the surface of the ground is not very great. There is one grave objection to high pile-trestles, and that is that the top end of the tree, and hence the poorest timber, is in the ground, and is liable to very rapid destruction by the elements at the ground-line. In order to retard this decay as much as possible, it is recommended in the Report of the Ohio Railway Commissioners for 1884 that the piles be painted for a short distance above and below the ground-line with hot tar. It has also been said that a coat of whitewash is beneficial where there is no water other than rain to wash it off.

The timber used for piles varies with the location, depending very largely upon the kind growing in the surrounding country. Among the varieties employed are the following, to be preferred in the order named, the first being the most durable:

Red Cedar.	White Pine.	Post Oak.
Red Cypress.	Redwood.	Red Oak.
Pitch Pine.	Elm.	Black Oak.
Yellow Pine (close-	Spruce.	Hemlock.
grained, long leaf).	White Oak.	Tamarac.

Order not known:

Red Ash.	Chestnut.	Buttonwood.
White Ash.	Beech.	Red or Norway Pine.
White Cedar.	Scrub Oak.	

They should be of straight, sound, live heart timber, perfectly free from windshakes, wanes; large, loose, black, or decayed knots; cracks, worm-holes, and all descriptions of decay; and should be stripped of bark. Some engineers prefer the piles to be hewed or sawed square. If piles are squared, they should be hewed rather than sawed, and be as free as possible from axe-marks. Squared piles ought to be at least 12 inches across each face, and not show more than 2 inches of sap at the corners.

Round piles are, as a rule, from 12 inches to 15 inches across the butt after being cut off, and when they are wider than the cap, the portion which projects on either side should be adzed off to an angle of at least 45° (Fig. 7).

The arrangement of the piling varies considerably, almost every constructor having his own plan and ideas. The nature and amount of the traffic should be carefully considered. For bents up to five feet in height, where the traffic is not very heavy, but three piles driven vertically are required. One should be placed on the centre-line, and one on either side from 3 feet 6 inches to 5 feet out (Fig. 3).

When the bents are from 5 feet to 10 feet high, and on lower ones on trunk lines, or where the traffic is heavy, four piles driven vertically should be used. The inner ones may be spaced from 4 feet to 5 feet between centres, and the outer ones about 11 feet from centre to centre (Fig. 4). If the piles can be driven into the ground for a depth of 8 feet or 10 feet, and have a good bearing, it will not in general be necessary to use sway-bracing.

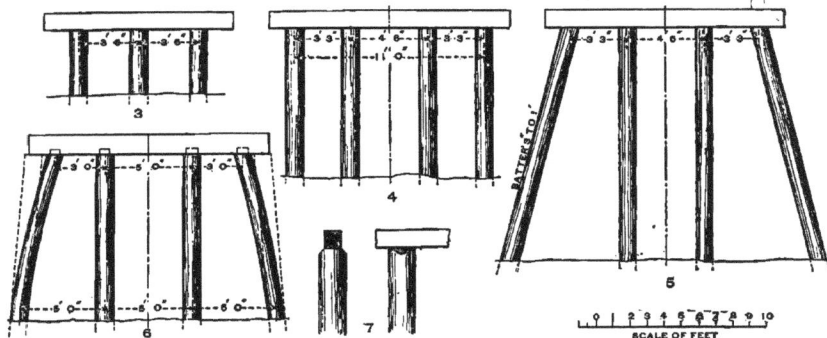

FIGS. 3 TO 7.—PILE-BENTS.

Above 10 feet in height it is well to drive the outside piles at a batter. According to present practice, this varies from 1 inch to 3 inches per foot. From 2½ to 3 inches is to be preferred, as it gives a broader base and greater stiffness to the structure (Fig. 5). The outer piles then perform to a certain extent the function of sway-braces and guys as well as supports. Bents between 10 feet and 20 feet high should be braced with one set of sway-braces, while above this it is advisable to divide the bent into two stories, so far as the bracing goes, making use of two X's, with two horizontal sticks between them. It is frequently well also to use longitudinal girts. But this subject of bracing will be thoroughly discussed in a succeeding chapter.

Instead of driving the outside piles at their full batter, the Burlington & Missouri River R. R. in Nebraska * drive them with a batter of 1 inch per foot, and then spring the top ends to place (Fig. 6). The following table gives the spacing:

TABLE VI.
Spacing of Piles, Burlington & Missouri River R. R. in Nebraska.

Height of Bent.	Inside Piles (vertical).		Outside Piles. Driven batter 1 in. per ft.	
	$c - c.$	Distance either side centre-line.	$c - c$ at ground.	Distance either side of centre-line.
Up to 10 feet, ...	5 ft.	2 ft. 6 in.	11 ft. †	5 ft. 6 in.
10 to 16 feet, ...	5 "	2 " 6 "	15 "	7 " 6 "
16 to 22 "	5 "	2 " 6 "	17 "	8 " 6 "
All at top immediately beneath cap, . .	5 ft.	2 ft. 6 in.	11 ft.	5 ft. 6 in.

* I. S. P. Weeks, Chief Engineer C., B. & Q. R. R., west of the Missouri River.

† Outside piles vertical.

Where soft ground extends to a great depth two or more piles may be fastened together end to end, if necessary The first pile is driven until the top is nearly to the surface of the ground or water, when it is cut off, trimmed up, and the second pile stood upon and fastened to it. The driving is then continued as before, and more piles added in the same way if required. The splice (Fig. 8) was used in the false-work for the erection of the Poughkeepsie Bridge, and is said to have proven very stiff and strong, and to have given great satisfaction.

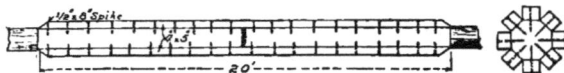

FIG. 8.—PILE-SPLICE, POUGHKEEPSIE BRIDGE.

Piles are also joined together by a long iron dowel (Fig. 9). The dowel is only of use to prevent lateral movement, and cannot be expected to keep the piles in line at all, on account of the great leverage. A wrought-iron dowel $1\frac{1}{2}$ inches in diameter by 2 feet long is of good proportions. It is also better to band the end of the larger pile with a wrought-iron ring to prevent its being split. A broad band encircling a portion of both piles (Fig. 10) is not very serviceable unless it be fastened so securely that it cannot move from one pile to the other, as unless this is done it is usually found to be wholly on either one of the two piles after the first few blows. If a few track spikes are driven into the piles above and below the ring, this movement will be

FIG. 9.   FIG. 10.
PILE -SPLICES.

prevented. The abutting end of the lower pile should be as large as practicable, and where the piles are of such timber as to require the wasting of a large part of the pile to secure a reasonable diameter, the contract should name a price for such wasted material. However, no waste should be paid for as such which can be used in any other place on the same contract, and that which is paid for should be considered as belonging to the company, and the contractor not be allowed to remove it unless he is willing to repurchase it.

It has sometimes been found, where the ground is very soft and runny, and it is difficult to drive the piles to a firm foundation, that if, after driving to a moderate depth, they are allowed to stand quiet for a day or so, the surrounding material will settle against them, and they will safely bear their load, being supported by the friction on their sides.

Previous to driving, the piles should have the end that is to penetrate the ground somewhat sharpened. The point shown in Fig. 11 is a very good shape.

Whenever the pile is likely to encounter logs, bowlders, or any material likely to split it, or to broom the point to an injurious degree, it is usual to protect it either by a cast- or a wrought-iron shoe. These shoes, however, are apt to strip off in driving.

FIG. 11.
PILE-
POINT.

Figs. 12 to 15 show some of the different forms of cast-iron shoes used, and Figs. 16 and 17 some of those of wrought-iron. The one in Fig. 14 is cast around a 1-inch square drift-bolt. This is preferable to the one in Fig. 12, in which the pin is of cast-iron as well as the point, as the pin is liable to break off just where it joins the point. Fig. 15 shows probably the best form of cast shoe. The dowel is a drift-bolt, as in Fig. 14; while in addition there is a recess about 2 inches deep, with walls from 1 inch to $1\frac{1}{2}$ inches thick, cast in the top of the point. The ring so formed not only helps to keep the shoe from being forced off laterally, and thus relieves the pin of some of the strain, but it also aids in preventing the pin from splitting the end of the pile in case much lateral force is exerted against the point. Fig. 16 shows a wrought-iron shoe. The point is of small size and has four straps

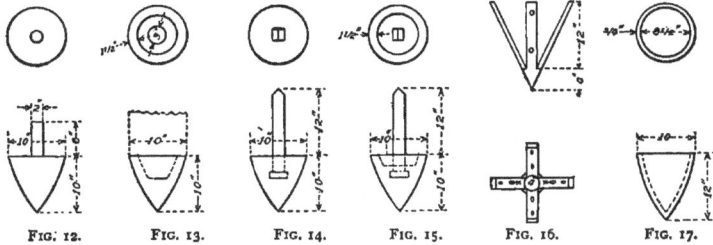

FIG. 12.   FIG. 13.   FIG. 14.   FIG. 15.   FIG. 16.   FIG. 17.

CAST- AND WROUGHT-IRON SHOES FOR PILES.

extending up the faces of the point of the pile, each of which is fastened to pile by two spikes.

Whenever the top of the pile is likely to be injured by the hammer it should be encircled by a heavy wrought-iron ring while being driven. Such a ring is shown in Fig. 18. These

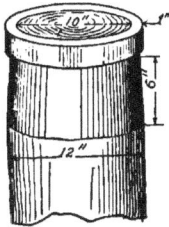

rings may be removed as soon as the driving is completed, and used over and over again. The temperature at the time of driving appears to have considerable influence over the tendency of the piles to split, and especially is this the case with certain kinds of timber, such as white and Norway pine. The colder the weather the greater the tendency. This is probably the case only when the temperature is below freezing. When the piles have been driven with the butts up in water, the elevation of the surface of which is liable to much change, the ice in the wintertime has been known to draw them in very cold climates.

FIG. 18.—RING FOR PILE.   In driving it is always preferable to have a heavy hammer with a short fall, but difficulties of transportation prevent the use of very heavy hammers in some cases. Ordinarily a hammer of 2000 lbs. weight falling 25 feet is a good proportion, but a heavier hammer with less fall is better. The piles should not settle more than from 1 inch to 3 inches at the last blow, and better if much less.

Where the bridge is very long, and hence the number of piles large, it becomes exceedingly important to determine with reasonable accuracy the approximate lengths of the piles and the number of each different length required. To accomplish this end a plan was adopted in the building of the Northern Pacific R.R. Bridge over the St. Louis River at Duluth, Minn., which is said to have proven very satisfactory indeed.* Test-piles were driven every 300 feet along the centre-line, and where any great difference in penetration was noticed an intermediate pile was driven. These piles were driven from a scow, and the distance measured by tying a rope to the last pile driven. The correct position of each pile was finally determined by triangulation. A complete record of the details of the test-piles was kept under the following form : †

---

* " Pile-driving," E. H. Beckler, *Eng. News*, vol. xvi., p. 83.

† This form has been very slightly modified by the author.

Sta.	No. of Pile.	Length.	Dia. of Top.	Diam. of Butt.	Depth of Water.	Dist. Driven.	Elev. Top of Pile.	Elev. Point of Pile.	No. of Blows.	Fall of Hammer.	Wt. of Hammer.	Penetration for each ten consecutive blows. 40	50	60	70	80	90	100	110	120	130	140	150
											lbs.												
80+23.7	16	52′	8½″	15½″	9.7′	39.3′	84.8′	32.8′	120	20′	2256			13″	14″	16″	17″	15″	12″	12″			

The fall of the hammer given is for the last ten, twenty, or thirty blows. Each pile was intended to be driven until it fulfilled the requirements of the specifications; i.e., penetration not to exceed 1 inch under the last blow of a 2000 lbs. hammer, with a fall of 20 feet or equivalent. The notes relating to the size and driving of the piles were taken by the inspector at the time; elevations by the engineer afterwards. A profile was then made up from these notes, and the piles ordered according to the lengths measured upon it. It was found that this method gave excellent results, there being but few discrepancies between the actual material when in place and that ordered. The waste of material was consequently reduced to a minimum.

A permanent record of all the final work, in detail, should always be made at the time construction is going on, for the future use of the Maintenance of Way-Department. An excellent form of such a record, as used on the above work, is given below.*

NORTHERN PACIFIC RAILROAD COMPANY.

PILE-DRIVING RECORD, BRIDGE NO. 160.

Date and Station.	Number of Bent.	Number of Pile.	Kind of Timber.	Length of Pile.	Diameter of Butt.	Diameter of Top.	Length below Cut-off.	Length of Cut-off.	Elevation of Ground.	Elev. of Point of Pile.	Distance driven.	Fall.	Number of Blows.	Penetration for given number of Blows.					
				ft.	ins.	ins.	ft.	ft.	ft.	ft.	ft.	ft.							
Dec. 27. 114+30	42	1	Wh. P.	40.0	17	14	31.4	8.6	76.2	66.4	9.8	20	85	20	10″ 40	10″ 60	5″ 70	4″ 80	1″ 85
114+30	42	4	Wh. P.	40.0	16	11	33.0	7.0	76.2	64.8	11.3	20	60	20	10″ 30	7″ 40	7″ 50	7″ 60	
Jan. 10. 112+65	53	3	Wh. P.	42.0	15	13	36.0	6.0	74.8	61.8	13.0	18	50			20	11″ 30	11″ 40	9″ 50

The actual cost and rate of driving the piles on this trestle was as follows:

SUMMARY OF WORK OF ONE DRIVER.

Dates.	Number of Days.	Number of Piles Driven.	Lin. ft. of Piles driven.	Number of Piles driven Daily.	Lin. ft. of Piles driven Daily.	Contractor's Cost of Driving.	Average Cost of Driving per lin. ft.
1884.							
Piles 23′ to 45′ long, Dec. 11 to 31.	10½	202	6520½	19¼	621	$454.43	$0.06969†
1885.							
Jan. 1 to 5 and 10 to 14. Piles over 45′ long.	5¾	134	5785	23 3/10	1006	212.57	0.0367
Jan. 5 to 10 and 14 to 31.	14½	364	21535	25 1/10	1485	539.61	0.02506
Feb. 1 to 28.	19¾	379	25036	19⅕	1268	747.74	0.0298
March 1 to 5.	3¼	73	4789	22½	1473	135.75	0.0284

Average cost per lin. ft., piles less than 45 feet long. $0.0542
" " " " " " over " 45 " " 0.0277

* This form has been very slightly modified by the author.

† Some delay on shore.

This is the actual cost of driving after the piles were delivered at the pile-driver.

The best record was 120 piles = 6600 ft. in four consecutive days. Contractors were Winston Bros. of Minneapolis.

This record may be taken as a fair average, and under like conditions may be used as a basis to estimate both time and cost. The trestle was a long one across a bay, which was completely frozen over.

In making up approximate estimates of the cost of piling and trestles, when exact data is not at hand, the following figures may be used with a very close approach to accuracy, as they are based on actual contract prices :

TABLE VII.

Prices of Trestle Material in Different Sections of North America.

Material.	Texas.	Virginia.		Indiana.	Washington.	Halifax Harbor.
	1888.	1889.	1890.	1879-81.	1888.	1884-85.
Average penetration of Piles on land.	*	†	‡	§ 12 ft.[2]	‖	¶
Piles per lin. ft. in place,	35c.			25c.		
Round Timber per lin. ft. erected (round-timber trestle)					9c.	
Dimension Timber per M, B. M., erected,	$40				$23	{ Hemlock, $16.47
Oak,		$30[1]	$30[1]	$25 to $30		
Pine,				$13 to $16		Wh. $26.00
Heart Pine (Va. Pine),		$30[1]	$30[1]			
Cattle-guard Timber per M, B. M., erected,	$30					
Bolts and Nuts per lb.,				5½c.	6c. to 7c.	

[1] Includes iron.

[2] This may be taken as the average penetration of piles in the clay of the Indiana prairies. On the Illinois prairies the average penetration may be taken at about 18 feet.

* C. A. Wilson.

† Maintenance of Way Dept., Norfolk & Western R. R.

‡ New River Plateau R. R., an extension of the N. & W. R. R.

§ I., D. & S. Ry., E. A. Hill. On Indiana section lumber is very scarce, and oak bridge-timber twenty to fifty feet long costs from $13.50 to $50 per M, B. M.

Not erected materials cost as follows, f.o.b.; Piles in ordinary lengths, and from 10 to 12 in. in diameter at the smaller end, from 13½c. to 15c. per lin. ft. Bolts 3c. and cast washers 2c. per lb.

‖ Vancouver, Klickitat & Yakima R. R., *Eng. News*, June 9, 1888.

¶ See *R. R. Gazette*, 1886, p. 242.

As soon as the pile-driving has advanced a short distance, the preparation of the top of the piles for the reception of the cap is begun. The elevation of the top of the pile is marked by a line on the face of one or two piles in each bent by the engineer.

FIG. 19.—MARKING PILES FOR CUTTING OFF.

A narrow plank having a straight upper edge, and long enough to extend entirely across the bent, is then nailed on each side of the piles (Fig. 19), and the top cut off level or cut in far enough to form the tenon. A cross-cut saw worked by two men is very convenient for this work.

There are several ways of fastening the caps to the piles,—by mortise and tenon, by drift-bolts, or by dowels. For solid caps, a tenon 3 ins. thick, 8 ins. wide, and 5 ins. long is a very good size. The edges around the top of the tenon should be chamfered (Fig. 20). When tenons are employed, it is customary to use wooden pins (treenails) for fastening the parts together. The pins may be of any tough hard wood. White oak and locust answer all the requirements very well. They ought to be from 1 in. to $1\frac{1}{2}$ in. in diameter, and slightly tapered—say $\frac{1}{8}$ in. to $\frac{1}{4}$ in. (Fig. 20). The hole in the tenon should be somewhat nearer the top of the pile than that in the cap is to the edge, so that the pin when driven in will draw the two parts tightly together. Bolts, $\frac{3}{4}$ in. in diameter, have been sometimes used in place of pins, but are not as desirable; in fact, their use should be discouraged. When drift-bolts or dowels are employed the top of the pile is cut off square. Dowels frequently extend through the cap; generally one, sometimes two, drift-bolts or dowels per pile are used; one is amply sufficient. Details of the different kinds of drift-bolts and of dowels are given in full in the chapter on iron details.

FIG. 20.—PILE TENON AND TREENAIL.

Sometimes a mortise-and-tenon joint is employed for the outer piles, with the inner piles cut off square, and drift-bolted as shown in Fig. 21.

There is still another method of fastening the caps to the piles, which is rapidly becoming the general practice, which is by the use of split caps. Instead of using a single piece of timber for the cap, two pieces, each half the size, are employed. For instance, a single 12-in. by 12-in. stick is replaced by two

FIG. 21.—FASTENING CAP TO PILES.

6-in. by 12-in. sticks. A tongue or tenon about 3 in. thick and the full width of the pile is formed on its top, and one of these pieces placed on either side and held in place by one, or better two, $\frac{3}{4}$-in. or 1-in. bolts passed through at each pile. The sticks should not be notched, and they should rest evenly on the edges formed on top of the piles. This form of cap is claimed to have many advantages, among which may be mentioned—

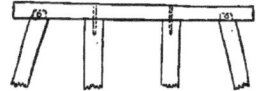

1st. On account of smaller size, better timber may be obtained and at less cost.

2d. Repairs may be made with ease and great economy in time and labor.

3d. Traffic need not be interfered with or endangered while repairs are being made.

4th. The caps may be replaced without cutting or injuring any other part of the structure in the least.

5th. Economy in material, because it is not necessary to replace the whole cap unless both sticks are decayed or injured, but only that part which is no longer in a serviceable condition.

CHAPTER III.

PILE-DRIVERS.

WHILE there are a great many forms and styles of pile-driving apparatus, there are but three principal methods of sinking piles in general use. These are:

1st. To force the piles into the ground by allowing a heavy weight to fall upon them when in an upright position, or by striking heavy blows by some means upon their heads or tops.

2d. To sink the piles by means of a jet of water.

3d. To screw the piles (which are either of iron or else have a special shoe) into the ground.

As the first of these methods is the one most extensively used, we may say almost universally,—and the one most generally applicable to trestle-building, we will confine ourselves strictly to a description of several different forms of apparatus for accomplishing the desired end by this means.

The particular kind of machine to be used will depend upon the special conditions surrounding the case.

In very rough and bad country the simpler and lighter the machines consistent with the requirements of the work the better. Sometimes merely a pair of leads with the necessary stays or back-bracing to give them the required stiffness, a common hoisting-machine (usually horse-power in such a case as this), with the hammer, ropes, and blocks, are all that are carried from place to place. Everything is made as simple as possible, and so that it can be taken apart for transportation. Sometimes the apparatus is mounted upon wheels, so that it may be folded down and drawn around by a team of horses. When the scene of action is reached the leads are merely raised up. This lifts the wheels off the ground. The base is then lashed to a couple of 12-in. logs, and as soon as the hoister is put in position and the tackle arranged, everything is ready for the commencement of the driving.

Where transportation is not too difficult, it is preferable to use a more complete driver. A steam-boiler and hoister is substituted for the horse-power one. With this arrangement the driving proceeds more rapidly and at less cost.

When many piles have to be driven in navigable waters the driver is mounted on a scow.

Figs. 22 to 28 illustrate a machine of the very latest model, and one of the heaviest in New York Harbor.*

"The hull is 56 ft. 6 in. long, and 23 ft. 6 in. wide, over all; each of the sides of the hull is made of four pieces of yellow pine, the two lower each 8 in. × 14 in., the third 7 in. × 14 in., the top piece 6 in. × 14 in., all securely tied by through-bolts; the bow planking is

---

* From "Bearing Piles," by Rudolph Hering, N. Y., 1887.

oak 5 in. thick, the bottom and end plank yellow pine 3 in. thick. The bow is further strengthened by a 16-in. × 16-in. cross-timber at top, and at the stern is an 8-in. × 12-in. cross-timber of yellow pine. Oak is used on the bow, as being better adapted to stand the

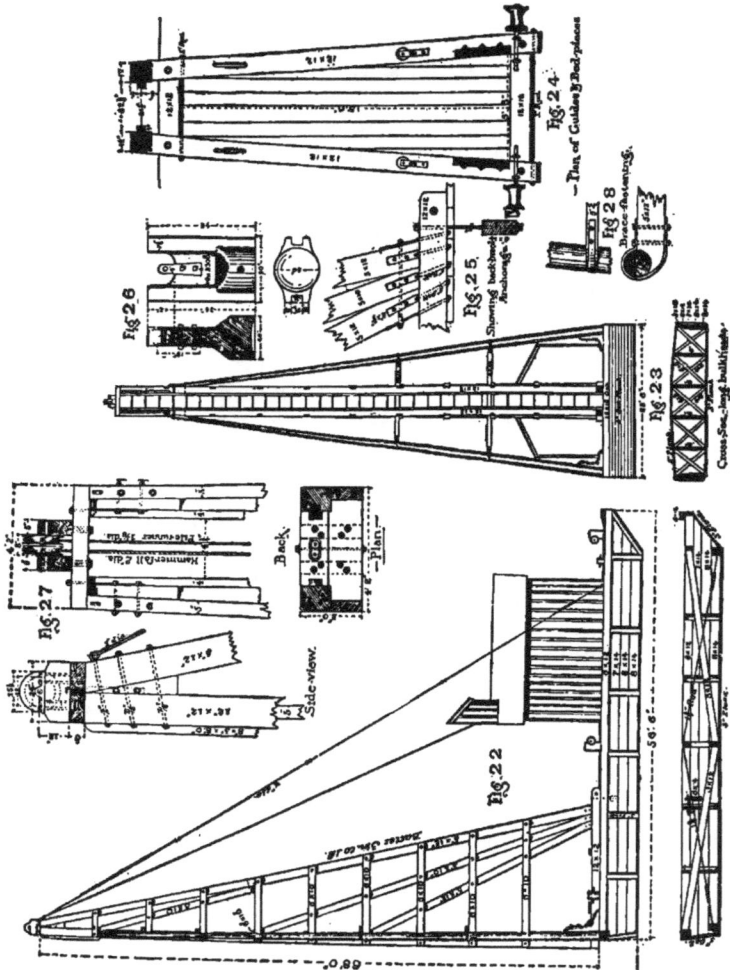

FIGS. 22-28.—FLOATING PILE-DRIVERS.

constant wear of the piles hauled against it; and to prevent knots or inequalities on the piles interfering with their position under the hammer, the bow planking overhangs 6 in. in its total height.

" The chief end in the design of a hull for a floating pile-driver is to obtain longitudinal stiffness, so that the strains between the bow and engine may be properly distributed. To this end our hull is strengthened lengthwise by four wooden bulkheads or kelsons, each 6 in. thick (Fig. 23), and braced laterally by four sets of X braces of 6-in. × 6-in. timber. The hull is further braced in the centre by two 3-in. × 12-in. Y. P. braces, and tie-rods or ' hog-chains' of iron 1¾ in. in diameter. Wale-pieces and fender-plank 3 in. thick protect the outside of the hull against chafing; .the deck has a ' crown ' of about 6 in. in its total width."

" The hammer-guides are made of two pieces of 12-in. × 12-in. Y. P. 67 ft. long from out to out, with inside guides of 5 in. × 4 in. stuff protected by plate-iron ¼ in. thick; ⅝-in. bolts with countersunk heads fasten the inner guides to the main sticks, and at the same time secure the iron-work to the same. The bottom of the main guides are connected with the 12-in. × 12 in. bed-pieces, shown in Fig. 24, by two timber-knees, and are tied at top by the cap shown in Fig. 27.

" The dimensions and general arrangement of the back-bracing is fully shown in Figs. 22 and 24; the bolts used in this portion of the framework are ⅞ in. diameter. The side-braces are round timbers 16 in. diameter at the butt, and they are anchored to the hull by two heavy timber-knees to each. The bed-pieces, as shown at Fig. 24, are fastened down to the hull by four bolts, each 1 in. in diameter, the forward bolts passing through the 16-in. × 16-in. oak piece on bow, and the after-bolts passing into a cross-timber 6 in. × 14 in., as shown at Fig. 25. The foot of the back-bracing is secured to the bed-timbers by one 1-in. strap-bolt in each timber, the strap portion of bolt being 2 in. × ¼ in. in section. A ⅞-in. through-bolt ties the three braces together.

" The iron stay-rods running from head of guides to after part of hull are two in number, and are each 1 in. in diameter.

" The hoisting-sheaves on top are two in number, placed side by side. They are 12 in. in working diameter, 15¼ in. from out to out, and 3½ in. wide; and the pin passing through them is 2½ in. diameter at the sheaves, and 2 in. diameter in the boxes. Experience teaches that these proportions are none too great to stand the severe work frequently put upon it in hoisting heavy weights and tearing out timber. The fall-rope attached to the hammer is 2 in. in *diameter*, and the ' runner' used in hoisting up piles is 1⅝ in. diameter.

" The hoisting-engine is a double-drummed Mundy engine of a nominal 25 horse-power.

" Fig. 26 shows the hammer used with this machine. The drawing is sufficient to show its general design. The weight is 3300 lbs.

" Fig. 28 shows the method of attaching the two 5-in. × 12-in. horizontal braces to the round side-braces, as further shown in Fig. 23."

In double-tracking a single-track road, or in repairing trestles in use, a form of driver mounted on a flat-car is found to be very convenient and economical. Figs. 29 to 32 show the details of one of the latest designs for a driver of this kind.*

It was constructed by the Missouri Pacific Railway, " with the purpose of obtaining a machine which could work effectively on piles at a further distance from the road-bed than usual. The design was worked out jointly by the Bridge and Building Department and the Car Department.

Railway Review, October 25th and November 8th, 1890.

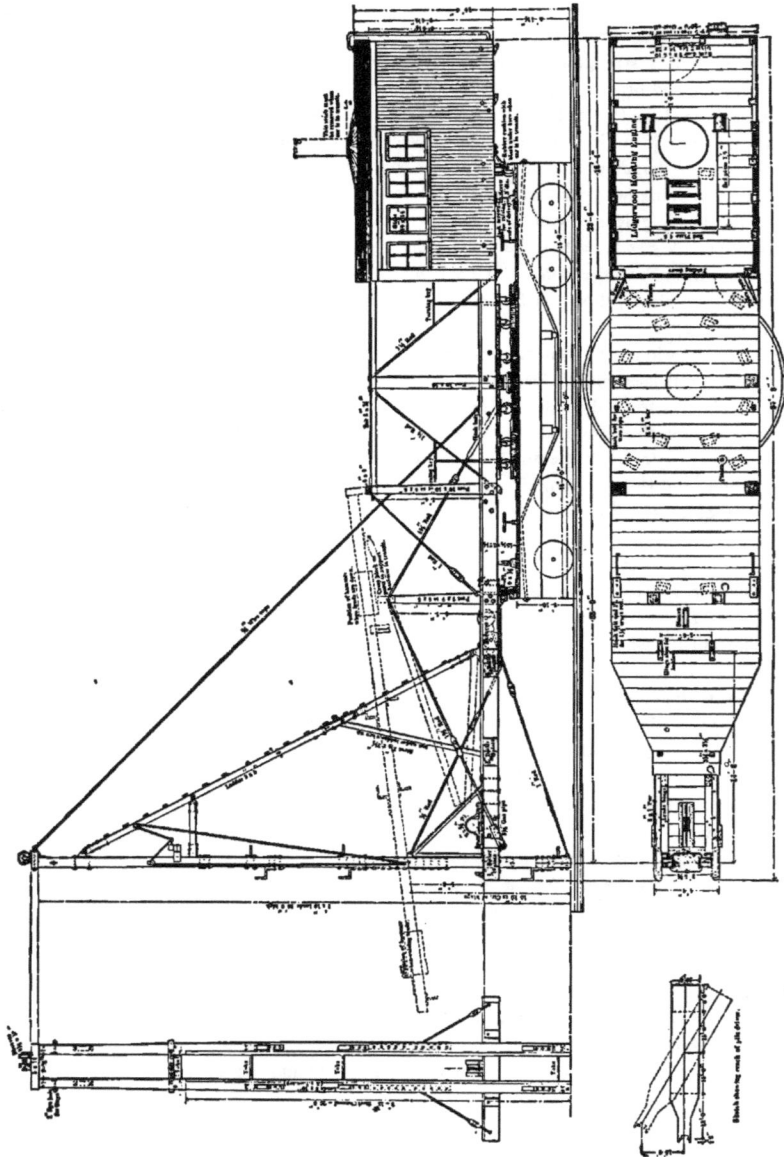

FIG. 29.—CAR PILE-DRIVER, MISSOURI PACIFIC RAILWAY.

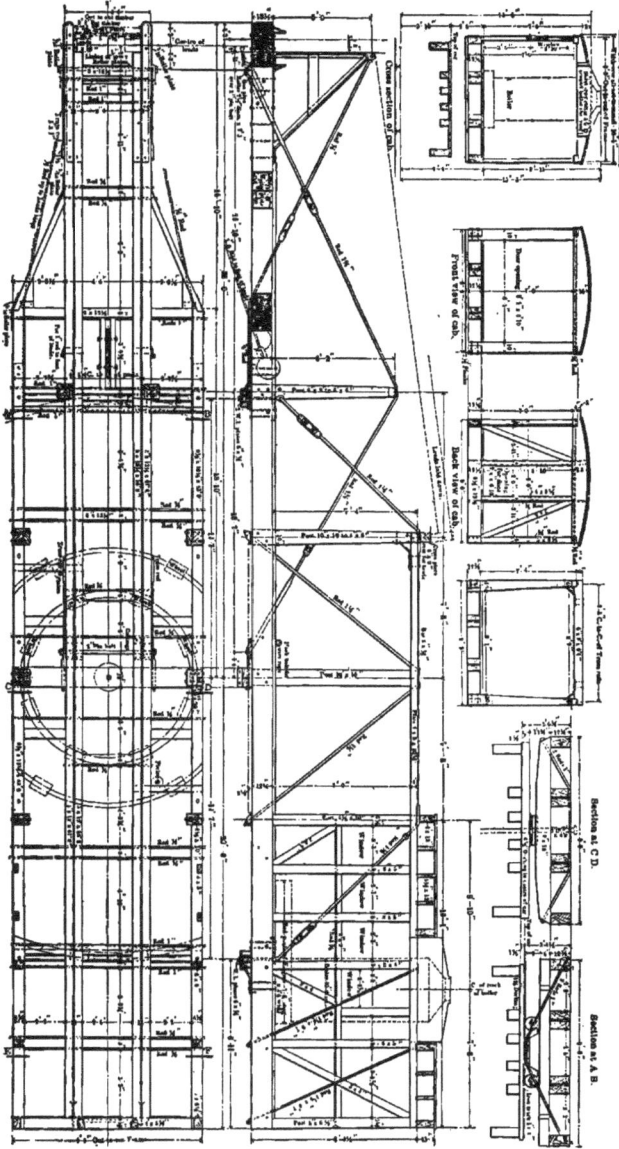

FIG. 30.—CAR PILE-DRIVER, MISSOURI PACIFIC RAILWAY. (FRAMING OF UPPER DECK AND CAB.)

" Fig. 30 shows the framing of the upper deck of the pile-driver and of the cab. It will be noticed that the main timbers are very long—57 ft. 8 in., and are 5 in. × 12½ in. in thickness. The side-sills are 6⅞ in. × 12¼ in., and 43 feet long. From the centre of the track on which the platform revolves to the centre of the leads is a distance of 33 ft., and in order to reach work which is located 16 ft. to one side of the centre of the track the driver must swing to an angle of about 30° from the track. The upper platform travels upon three circular tracks. The first is a complete circle, having a diameter of nearly 9 ft., and as the car is 9 ft. wide and the upper platform 10 ft., this track is fixed. The next circle has a diameter of 13 ft. 3 in., and is composed of four pieces of rail of the ordinary section, two of which are firmly secured to the car platform, while the other two pieces overhang the sides of the car, and are removed while the pile-driver is in transit. When in use they are supported in position by two wrought-iron swing-brackets fastened to the outside face of each side-sill, and are also secured to fixed sections by fish-plates. The third circular track has a radius of 14 ft. 5 in., and is a bar of iron 4 × 1 in. This track is not carried beyond the sides of the car. The wheels which bear upon the two smaller circles are attached directly to a heavy flooring on the under side of the platform, and as far as possible they are placed in the vicinity of a longitudinal sill, so as to give them as solid a bearing as possible. The rollers which bear on the outer one of the three tracks are secured to the under side of a heavy transverse bolster, which is composed of three pieces of wood with three wrought-iron plates 6 in. × ⅞ in. in section intervening between them, the bolster being 6 in. × 10½ in. in section. The bolster at the centre-pin is wood 12 in. wide and 9 in. deep, and is trussed by two rods each an inch in diameter.

" The construction of the leads and ladder will be best understood by a reference to Fig. 29. The leads are 36 ft. long, and are hinged to a heavy triangular framework, a detail of which is shown in Fig. 31. A sole-plate 10 in. × ⅜ in. in section is secured to the upper face

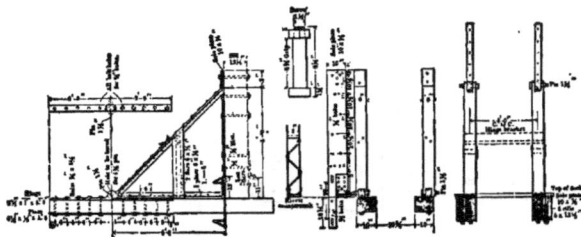

FIG. 31.—CAR PILE-DRIVER, MISSOURI PACIFIC RY. (DETAILS OF LEADS.)

of the longitudinal sills which support the beam. The hinge-frame is secured to this plate, and consists of plates 6 in. × ⅞ in., reinforced by angle-irons. The inner faces of the leads are protected by steel channel-irons extending up from the bottom end for a distance of 26 ft. These afford a good bearing for the hammer, which is planed out to fit them.

" The car upon which the pile-driver is carried is shown in Fig. 32. It is an exceedingly heavy car, as will be seen from an inspection of the drawing. It is 30 ft. long and 9 ft. wide, and very strongly trussed. The rack for moving the upper deck is seen in this view, and requires no explanation. When the car is in transit, four jack-screws, one at each corner of the car, are adjusted against suitable sockets on the under side of the upper deck, so as to steady the entire superstructure. At the same time the upper deck is prevented from swinging out of a longitudinal position by means of suitable hooks attached to the upper deck, which engage eye-bolts in the ends of the car. When the pile-driver is at work these jack-screws are released, and the heavy screws seen extending down through the floor of the car are made to bear upon the truck frames. This prevents any undue strain on the trucks springs, and also any unsteadiness which might be caused from their elasticity."

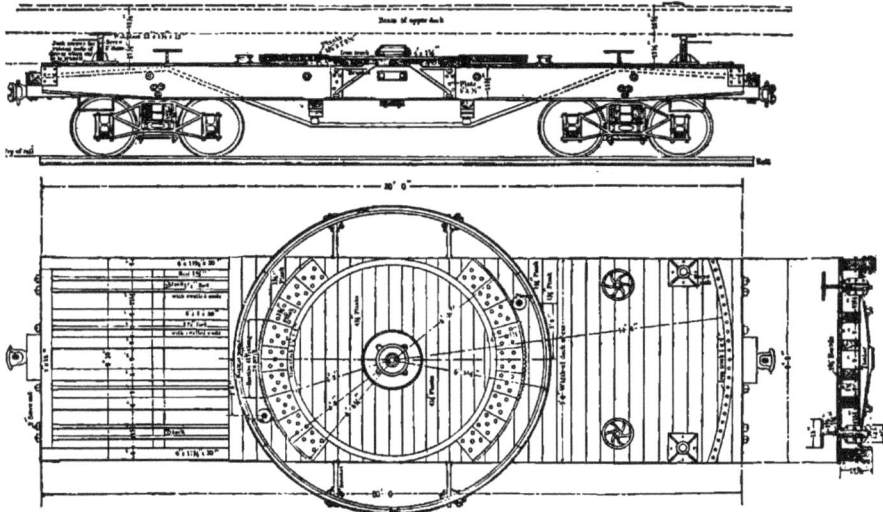

FIG. 32.—CAR PILE-DRIVER, MISSOURI PACIFIC RY. (DETAILS OF CAR.)

" The hammer weighs 2397 lbs., and is operated by a Lidgerwood hoisting-engine. As will be seen from the illustrations, the upper platform of the pile-driver is so long in comparison with the car which carries it that a flat-car at each end is necessary for its transportation. One of these cars is constructed for the purpose, and carries a supply of water and coal, and any other material which may be necessary. This is attached at the cab end of the pile-driver, and the one at the other end is a common flat-car."

In use the ordinary drop-hammer tends to batter the heads and split the piles. It is claimed that the steam pile-hammer overcomes this objection to a very large extent, and that it will drive the piles more rapidly.

"After the steam-hammer begins operation the blows are so rapid—70 to 100 per minute—that the earth once disturbed has no time to settle, and the pile sinks through it in somewhat the same manner as it would when the earth is loosened and held in suspension by a water-jet." *

* " Builders and Contractors' Engineering and Plant :" *The Sanitary Engineer.*

Messrs. Ross & Sanford drove on an average 83 piles, having a penetration of 17 ft., in material mostly sand and oyster-shells, per day of ten hours, with one of these steam-hammers. There were 1459 piles in all in the work, which was a pile-dike in the Passaic River, N. J. The best ten-hours' work was 121 piles. The best work in the same time with an ordinary driver was 63 piles. In the matter of expense the steam-hammer costs more for steam than the ordinary driver, but this is more than offset by the saving in pile-bands, rope, and the number of men on the machine.

FIG. 33.—STEAM PILE-HAMMER IN OPERATION.

Figs. 33 to 37 show such a hammer, and the method of using it. The following reference-letters will aid in enabling one to understand the construction of the machine :

A A. Ten-inch I-beams forming the sliding-frame within which the hammer H L slides.

B. Cross-girder riveted to the upper ends of the I-beams (A) by means of which and the bail (S) the whole apparatus is raised or lowered for adjustment to the head of the pile.

C. Hollow piston-rod hung loosely on a collar through a hole in the cross-girder (B).

D. The steam-chest, supported by the cross-girder (B) and covering the opening of the piston-rod (C).

E. Piston-head and plug of the end of the hollow piston-rod (C).

F. Steam-openings in the hollow piston-rod (C) through which steam passes to the space (G) surrounding the rod.

G. Annular space (between the piston-rod and the interior of the hammer) which forms the steam-cylinder.

H. Hammer-cylinder.

I. Cylinder-cover with stuffing-box.

J. Foot-block or bonnet casting riveted to the lower ends of the I-beams and forming the lower part of the sliding-frame.

K. Conical opening through the foot-block (J) shaped to receive the head of the pile.

L. Cylindrical prolongation of the hammer-cylinder (H), forming the hammer-head.

M. Lever which works the steam-valve.

N. Fulcrum for the lever (M) bolted to the face of the girder (B).

O. Attachment of the lever to the valve-stem.

P. Upper trip, which throws the valve-stem in and supplies steam to the cylinder (G).

Q. Lower trip, which throws the valve-stem out and exhausts the steam. (P and Q are both attached to the hammer-cylinder.)

R. Vent for air and condensed water.

S. Bail by which the whole machine is lifted.

T. Connection of steam-hose with steam-chest.

FIG. 34.

FIG. 35.

FIG. 36.

End-projection, Valve A.

Side-projection, Valve B.

Clamps D.

FIG. 37.—BALANCE-VALVE.

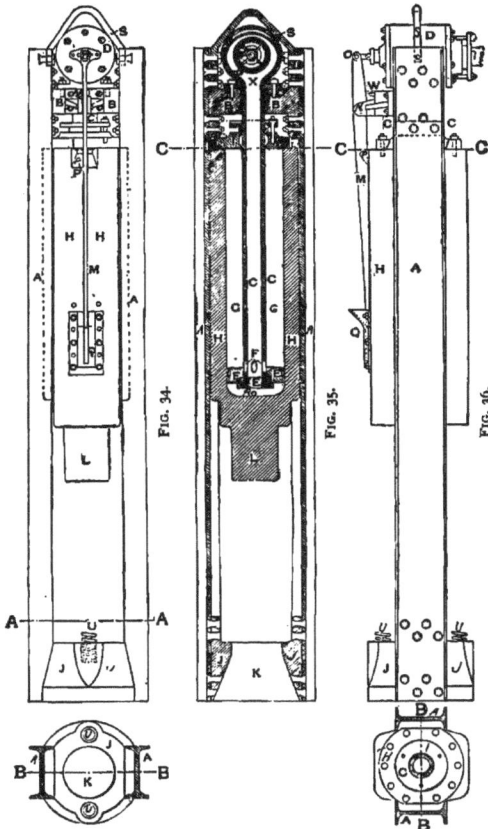

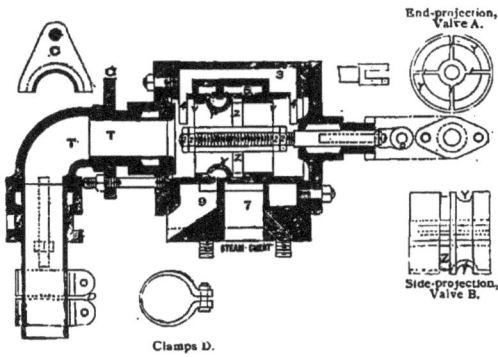

FIGS. 34 TO 37.—DETAILS OF CRAM'S STEAM PILE-HAMMER.

U. Springs to guard foot-block.

V. Balanced steam-valve.

W. Set-screw regulating travel of steam-valve.

X. Mouth of steam-chest (D).

" Fig. 33 is a part general view of a scow pile-driver, with the machine lowered and resting on a pile.

" Fig. 34 is a rear elevation of the machine, showing the valve-tripping apparatus, also a sectional plan on line **A A**.

" Fig. 35 is a central vertical section on line **B B**.

" Fig. 36 is a side view, also section on **C C**.

" In use the whole apparatus slides on a pair of ordinary ways or 'leaders,' being raised or lowered by means of the bail (S).

" When ready for use, as shown in Fig. 33, the hammer is at the bottom of the frame resting on the springs U,—or, if the pile is set up, on the pile,—the piston-head being at the *top* of the cylinder.* These springs U serve to catch any chance blow and prevent injury to the foot-block (J). The hammer-head, of course, projects through the orifice (K) in the foot-block, and as much as the pile-head enters the orifice, so much is the hammer-cylinder pushed up, so that its whole weight rests on the pile.

" On the admission of steam through the flexible hose attached to the inlet (T), it passes through the piston-rod into the cylinder (G), the hammer slides upwards in its frame A B J to the extent of its stroke (or about 40 inches for a large machine); then the steam is exhausted through the tripping of the valve, and the hammer falls, giving a free blow. The upper trip at once admits steam, and the operation is quickly repeated."

" Fig. 37 shows details of the valve and steam-chest.

" The valve itself is shown in end projection at A, and side projection at B, and is a hollow cylinder, with open ends and a ring Y cast around its periphery, with a slot Z cut through its shell near the ring, and with the socket by which it is held upon the valve-stem supported by four radial webs (1) extending the length of the valve and attached to its shell. The jamb-nuts (2) hold it firmly upon the valve-stem, where the valve is shown in place in section of steam-chest between lines V V. The upper wall of the steam-chest at (3) is made hollow to preserve equal thickness of metal for uniform expansion, etc., and is connected by openings (4) with interior of steam-chest, so that steam finds constant admission and can circulate from end to end of steam-chest in this way as well as through the hollow valve.

" The cylindrical box (5) is cast so as to surround the valve, and connects, first, through slot 6 with the opening 7, which joins the hollow piston-rod leading to the hammer cylinder, and, second, with the exhaust orifice 8 through 9, which opens to the air. The tongue (10) separates the box into two portions used for exhaust and supply respectively, each of which has an annular slot (8 and 6) through the steam-chest shell surrounding the valve shell.

" It will be noticed that the only steam-pressure on the valve is outward from within the cylindrical valve-shell, and can exert no influence toward producing friction. In other words, the valve is so perfectly balanced that it works readily, whether with or without steam supplied.

---

* Corrected from original.

" In the drawing the valve is slid out, or as it would be thrown by the trip Q acting upon the lever M when it is ready to fall.   In other words, the steam is being exhausted as the ring Y straddles the two slots 6 and 8, and steam from within the hammer cylinder finds vent up the hollow piston-rod and out the exhaust-port (9).

" The complete action is then: The hammer falling, the valve is thrown so the slot Z covers 6, while the exhaust-slot 8 is covered by metal.   The steam from steam-hose connection T through the hollow valve-slot, Z, 6 and 7, and hollow piston-rod finds admission to the interior of the hammer, which causes it to rise until the lever throws out the valve, and steam is exhausted as before.

" The hammer-cylinder weighs 5500 lbs., and with 60 to 75 lbs. steam gives 75 to 80 blows per minute.   With 41 blows a large unpointed pile was driven 35 feet into a hard-clay bottom in half a minute.

" The steam-valve has a travel of five eighths of an inch in a steam-jacketed chest.   The length of its movement is adjustable, so as to suit the force of the blows to the work in hand."

The cost of these drivers varies according to size and weight of ram, as follows :

Size.	Weight of Ram.	Price.	Width between Leaders.	Length of Stroke.	Total Weight.	Total Length.
B	5500 lbs.	$800	27 inches	40 inches	8400 lbs.	12 feet
C	3000 "	700	20 "	40 "	5500 "	12 "
D	2000 "	500	20 "	24 "	4200 "	8½ "

Including 30 feet of steam-hose and couplings.

CHAPTER IV.

FRAMED BENTS.

FRAMED BENTS are built upon a foundation of some kind, the object of which is to raise the sill from the ground and thus lengthen its life—which at the best is short enough—as much as possible. When the sill is partly or wholly buried in the ground decay soon sets in and proceeds with great rapidity, and the practice either of allowing the sill to rest upon the ground or of partially or wholly covering it with earth is to be very strongly condemned.

The foundations may be divided into seven classes: masonry, pile, mud-sill or sub-sill, grillage, crib, solid rock, and loose rock.

Masonry foundations are of third-class masonry, and are built of such material as may be found near at hand. The stones should be as large and flat as possible, all those with any rounded surfaces being carefully excluded. The masonry should penetrate the ground to below the frost-line so as to prevent heaving, which would tend to rapid disintegration and destruction. It should also rest upon a firm bed. For low trestles, where the sills are short, the masonry may extend the whole length of the sill, but where the bents are over 10 ft. in height it becomes more economical to divide it into three sections, placing one part under the centre-posts and one under either batter-post. The part under the centre-posts should be long enough to give a good solid bearing to each of them, and to extend some little distance beyond them. The faces of the foundation should have a slight batter, thus giving it. the shape of a truncated pyramid. Figs. 38 and 39 show the shape, size, and arrangement of the masonry foundations as used on the New York, Ontario & Western Railroad.

FIG. 38. FIG. 39.

MASONRY FOOTINGS.

Masonry forms excellent foundations—more durable than any other kind. They are practically indestructible if well built, not being liable to decay, and are very economical in maintenance, as the life of the sills is greatly prolonged, and the repairs to the foundation amount to practically nothing. It is well (but hardly the general rule) to fasten the sills to the foundations by means of iron rods built into the masonry. This prevents vibration to a very considerable extent.

These rods should be from $\frac{3}{4}$ in. to 1 in. in diameter and about 3 ft. long. If desired, a head may be formed on the lower end and anchor-plates employed, though this is not essential.

In pile-foundations one pile is usually placed under each post. The sill is fastened to

24

the piles in any of the ways previously described for fastening on the caps of pile-bents. In very high trestles it is at times found desirable to place two piles under each post. It is often convenient, especially where pile-timber is plenty, and desirable framing timber of a considerable length is difficult to obtain, to use a pile-bent of from 10 to 20 feet high, and then place a framed one on top of it. On some roads when the trestle is over a water-way, a pile-foundation is surmounted by a framed bent, the piles being of such a height as to always remain entirely submerged. By this means the decay due to the alternate wetting and drying of the timber is almost wholly confined to the framed portion, which is easily replaced: hence it is said that this style of structure is very economical over bays and inlets affected by the tides.

The practice as to mud-sills or sub-sills varies a great deal. Some prefer thin planks only 3 in. or 4 in. thick; others, material 12 in. by 12 in. The thicker material is the better, as it raises the sill higher from the ground, and is not so rapidly weakened by decay. Sub-sills should only be used when the surface is rock or where the ground is quite hard. In

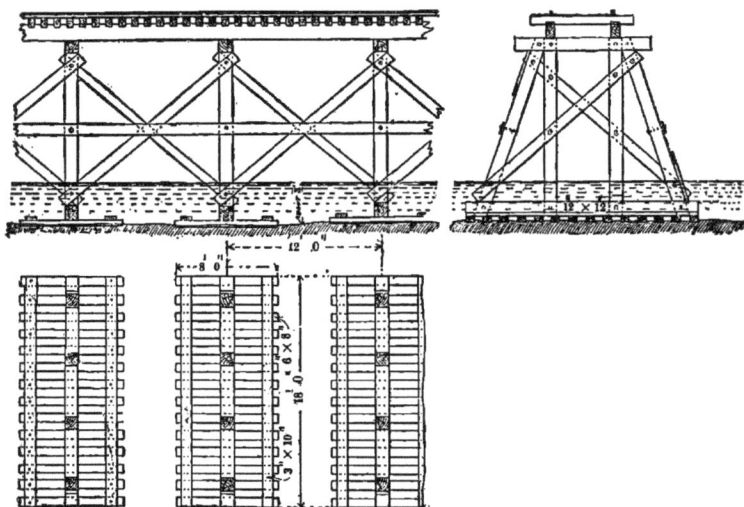

FIG. 40.—GRILLAGE FOUNDATION.

other cases piles or masonry should be employed, depending on circumstances. Some go to the trouble of notching the sub-sill and sill together. This is not necessary, and adds more to the expense of construction than the good derived from it warrants. It is always better to spike or drift-bolt the two together to some extent, though it is not necessary to fasten every sub-sill in this way when there are more than four under the bent. When longitudinal girts are placed immediately above the sill, no fastening at all of the sill to the sub-sill is required.

A grillage was used as a foundation for a trestle on a branch of the New York, New Haven & Hartford Railroad, over Hanover Pond, at Meriden, Conn. The water was still and shallow, and the bottom soft, treacherous, and of unknown depth. Piles could not be economically used, and hence Mr. J. Devin devised this expedient.

The grillage was made by arranging a number of railroad ties about 15 in. apart, as shown in Fig. 40, and fastening them together by two 3 in. by 10 in. binder-planks spiked to each tie. It was then floated to its place and a framed bent placed upon it, one grating being used for each bent. The weight of the structure sunk the grillages so that they rested on the bottom. The sills were not fastened to them, the bents being kept in place by both sway and longitudinal bracing. It seems as though it might be more advisable ordinarily to fasten the sills to the foundation by a few drift-bolts. This foundation is said to give perfect satisfaction under constant traffic. One grating was put in with the binder-planks underneath, but was quickly undermined its whole length either by a slight current or by springs. The trouble was obviated by turning the grating over.*

On some of the branches from the Cripple Creek extension of the Norfolk & Western Railroad, crib foundations are used for the trestles. These cribs are formed by piling logs on top of each other, notching them where they cross, and then filling up the interior with stones (Fig. 41).

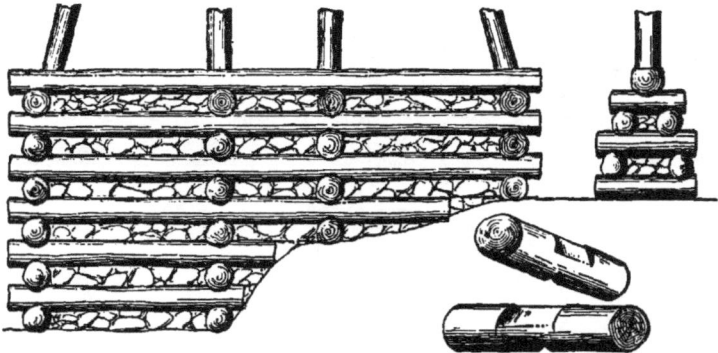

FIG. 41.—CRIB FOUNDATIONS.

They are built pyramidal in form, and are suitable for side-hill work where the slope is not too great, though their use is not by any means limited to this kind of ground, as they form as good a foundation on the level. When on a side-hill the ground beneath them is excavated in steps, and the cribs are built up level so as not to necessitate the breaking of the sill. The logs composing the crib should be at least 10 in. in diameter at the smallest place, and it is better if they are not under a foot.

A novel plan for obtaining a foundation on a side-hill where the surface was of solid rock was adopted on some work by Mr. J. E. Woods, C.E. Holes were cut in the rock where the feet of the posts were to come, and after the posts were placed in them and the bent

* *Eng. and Build. Rec.*, Dec. 24, 1887.

completed, the remaining spaces were filled with cement (Fig. 42). Of course no sills are required with this form of foundation. The ends of all the posts were tarred before they were placed in the holes.

Where loose rock is plentiful, and cement or lime costly, pretty serviceable foundations may be obtained by filling trenches with it. Sub-sills should be laid on top of these, and the sills rest upon them. This is done in order to distribute the weight over a larger surface.

The life of sub-sills can often be greatly lengthened, and loose rock foundations kept

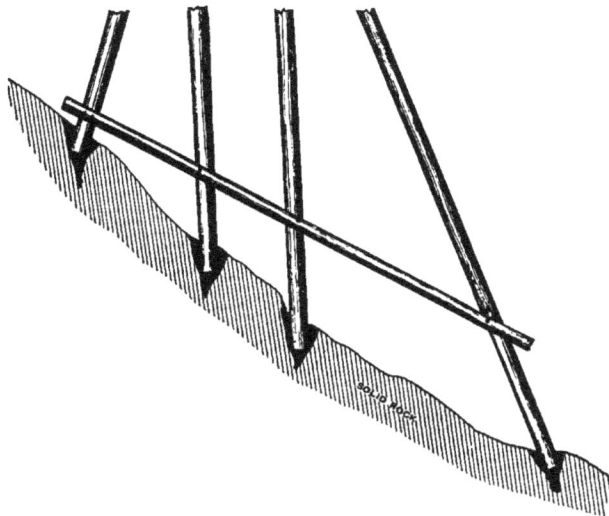

FIG. 42.—SOLID ROCK FOUNDATIONS ON SIDE-HILL.

quite dry, by digging a trench entirely around them, several feet away, and leading off the water that accumulates.

Sills (except when split) should not be of smaller timber than 12 in. square, and should extend from 12 in. to 18 in. beyond the outside of the batter-posts. In very high trestles the sills are usually made up of several pieces. Some examples of these are shown in the cuts of special trestles. When the sills are mortised, a drip-hole ¼ in. in diameter should always be bored with a downward inclination from the bottom of the mortise to the outside of the sill. Figs. 43 and 44 show the method of boring these drips on two roads.* The object is to keep the water from collecting in the mortises, which would hasten the decay of the sill.

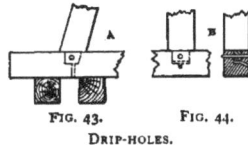

FIG. 43.   FIG. 44.
DRIP-HOLES.

---

* Charleston, Cincinnati & Chicago Railroad ; Central Railroad of Georgia, V. H. Kriegshaber, Assistant Engineer.

There are usually four posts to a bent—two vertical or plumb posts, and two batter-posts. As a rule, they are all made of the same size timber,—12 in. × 12 in. Occasionally either all, or else the batter-posts, are made of 9-in. × 12-in. or 10-in. × 12-in. material, and sometimes, in very low trestles, 10-in. × 10-in. is employed.

The large size, 12-in. × 12-in., is rather to be preferred in all cases, it being far more advisable to have an excess of strength in this part than any tendency to weakness. The extra cost for timber does not amount to very much. The plumb-posts should be spaced from 4 ft. to 5 ft. between centres, and the batter posts 11 ft. from centre to centre at the top, immediately under the cap. The inclined posts should have a batter of 3 in. per foot. This give a broad base, and adds considerably to the stiffness of the bent. Other batters are frequently made use of, varying from 2 in. to 4 in. per foot, though 2½ in. and 3 in. are the most common. Table VIII gives the length of the batter-posts for different heights at an inclination of 3 inches per foot.

TABLE VIII.

Length of Batter-posts; Batter 3″ per Foot.

Distance between Cap and Sill.		Length of Stick.					
		Shoulder to Shoulder.		Stick with Square Ends.		With two 5″ Tenons.	
ft.	in.	ft.	in.	ft.	in.	ft.	in.
3	0	3	1	3	4	4	2
3	6	3	7⅛	3	10⅛	4	8⅛
4		4	1½	4	4½	5	2½
4	6	4	7⅝	4	10⅝	5	8¼
5		5	1⅞	5	4⅞	6	2⅞
5	6	5	8	5	11	6	9
6		6	2¼	6	5¼	7	3¼
6	6	6	8⅝	6	11⅝	7	9⅝
7		7	2½	7	5½	8	3½
7	6	7	8¾	7	11¾	8	9¾
8		8	3	8	6	9	4
8	6	8	9⅛	9	0⅛	9	10⅛
9		9	3⅝	9	6⅝	10	4⅝
9	6	9	9½	10	0½	10	10½
10		10	3⅝	10	6⅝	11	4⅝
10	6	10	9⅞	11	0⅞	11	10⅞
11		11	4	11	7	12	5
11	6	11	10¼	12	1¼	12	11¼
12		12	4½	12	7½	13	5½
12	6	12	10⅝	13	1⅝	13	11⅝

Distance between Cap and Sill.		Length of Stick.					
		Shoulder to Shoulder.		Stick with Square Ends.		With two 5″ Tenons.	
ft.	in.	ft.	in.	ft.	in.	ft.	in.
13		13	4¾	13	7¾	14	5¾
13	6	13	11	14	2	15	
14		14	5⅛	14	8⅛	15	6⅛
14	6	14	11⅜	15	2⅜	16	0⅜
15		15	5½	15	8½	16	6½
15	6	15	11¾	16	2¾	17	0¾
16		16	5⅞	16	8⅞	17	6⅞
16	6	17	0	17	3	18	1
17		17	6¼	17	9¼	18	7¼
17	6	18	0½	18	3½	19	1½
18		18	6⅝	18	9⅝	19	7⅝
18	6	19	0⅞	19	3⅞	20	1⅞
19		19	7	19	10	20	8
19	6	20	1¼	20	4¼	21	2¼
20		20	7⅝	20	10⅝	21	8⅝
20	6	21	1½	21	4½	22	2½
21		21	7¾	21	10¾	22	8¾
21	6	22	1⅞	22	4⅞	23	2⅞
22		22	8⅛	22	11⅛	23	9⅛
22	6	22	2¼	23	5¼	24	3¼

Distance between Cap and Sill.		Length of Stick.					
		Shoulder to Shoulder.		Stick with Square Ends.		With two 5″ Tenons.	
ft.	in.	ft.	in.	ft.	in.	ft.	in.
23		23	8½	23	11½	24	9½
23	6	24	2⅝	24	5⅝	25	3⅝
24		24	8⅞	24	11⅞	25	9⅞
24	6	25	3	25	6	26	4
25		25	9¼	26	0¼	26	10¼
25	6	26	3⅜	26	6⅜	27	4⅜
26		26	9⅝	27	0⅝	27	10⅝
26	6	27	3¾	27	6¾	28	4¾
27		27	10	28	1	28	11
27	6	28	4⅛	28	7⅛	29	5⅛
28		28	10⅜	29	1⅜	29	11⅜
28	6	29	4½	29	7½	30	5½
29		29	10¾	30	1¾	30	11¾
29	6	30	4⅞	30	7⅞	31	5⅞
30		30	11	31	2	32	0
30	6	31	5¼	31	8¼	32	6¼
31		31	11½	32	2½	33	0½
31	6	32	5⅝	32	8⅝	33	6⅝
32		32	11⅞	33	2⅞	34	0⅞
32	6	33	6	33	9	34	7

Distance between Cap and Sill.		Length of Stick.					
		Shoulder to Shoulder.		Stick with Square Ends.		With two 5″ Tenons.	
ft.	in.	ft.	in.	ft.	in.	ft.	in.
33		34	0¼	34	3¼	35	1¼
33	6	34	6⅜	34	9⅜	35	7⅜
34		35	0½	35	3½	36	1½
34	6	35	6¾	35	9¾	36	7¾
35		36	0⅞	36	3⅞	37	1⅞
35	6	36	7⅛	36	10⅛	37	8⅛
36		37	1¼	37	4¼	38	2¼
36	6	37	7½	37	10½	38	8½
37		38	1⅝	38	4⅝	39	2⅝
37	6	38	7⅞	38	10⅞	39	8⅞
38		39	2	39	5	40	3
38	6	39	8¼	39	11¼	40	9¼
39		40	2⅜	40	5⅜	41	3⅜
39	6	40	8⅝	40	11⅝	41	9⅝
40		41	2¾	41	5¾	42	3¾
40	6	41	9	42	0	42	10
41		42	3⅛	42	6⅛	43	4⅛
41	6	42	9⅜	43	0⅜	43	10⅜
42		43	3½	43	6½	44	4½
42	6	43	9¾	44	0¾	44	10¾

The second columns in the table give the length of the post without tenons, measuring along one of the faces after the ends have been cut off at the proper angle; the third columns, the length of a piece of timber with square ends required to cut the post; and the fourth columns, the length of a piece of timber with square ends required when there is a tenon 5 inches long on each end. The table is used thus: What is the length of timber required for the batter-posts of a bent 21 ft. 6 in. high, the posts being connected to both cap and sill by a 5-inch tenon? Taking the thickness of both cap and sill from the height of the bent in order to find the distance between them, we have 21 ft. 6 in. — 2 ft. = 19 ft. 6 in.

Now looking in the table we find in the fourth column, opposite 19 ft. 6 in., that the length required is 21 ft. 2¼ in.

For framing in the field, try-squares, set to the proper angle for cutting the ends of the batter-posts, are very convenient. Fig. 45 shows a form of template for direct use. It consists of a ½-in. board cut to the requisite angle with a 1½-in. square piece fastened along one edge. It is used in the same manner as an ordinary carpenter's square.

Some designers prefer to have the batter-posts touch the plumb-posts where they meet the cap, as in Fig. 46, while others incline all of the posts (Fig. 47). When all of the posts are

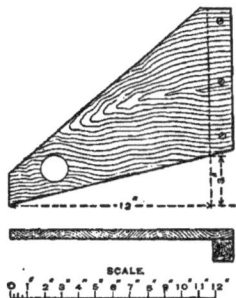

FIG. 46.

FIG. 45.
BATTER-POST TEMPLATE.

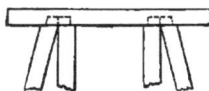

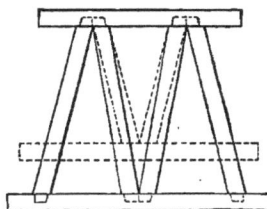

FIG. 47.
ARRANGEMENT OF POSTS.

inclined, the distance between them at the top is fixed, as is also the batter of the outer posts, while that of the inner ones varies with the height.

It is well to make solid caps of at least 12 in. × 12 in. timber and 14 ft. long. Where the timber is inclined to be weak or brittle, they should be 12 in. wide by 14 in. deep. There are six different ways of joining the sills, posts, and caps together, viz., by

| Mortise and tenon; | Dowels; | Iron-joint plates; |
| Drift-bolts; | Plasters; | Split caps and sills. |

A tenon 3 in. thick, 8 in. wide, and 5 in. long is a very good size. The mortise should be a little deeper—say ¼ in.—than the length of the tenon. They should be snugly fitted to

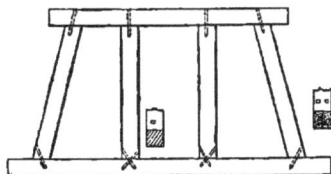

FIG. 48.—ARRANGEMENT OF DRIFT-BOLTS.

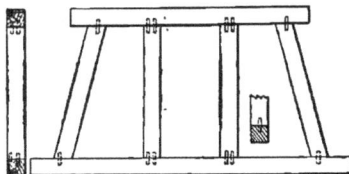

FIG. 49.—ARRANGEMENT OF DOWELS.

each other, and the sides made as smooth as practicable. The same precaution in regard to boring the holes in the tenons, as mentioned when speaking of the tenons on piles, should be observed here, so that the work may be drawn tightly together.   Wooden pins should always be used to hold the parts together.

When drift-bolts are employed, two should be used for fastening each post to the sill, and one for securing it to the cap.   A hole very nearly the size of the drift-bolt should be bored through the first stick of timber penetrated, and one somewhat smaller through the balance.   The drift-bolts may be arranged as in Fig. 48.

In dowel-joints two dowels should be used in both cap and sill to each post.   They should be ¾ in. in diameter, by at least 8 in. long, and arranged as in Fig. 49.

A plaster-joint is one of the most convenient forms for some uses.   It is especially advan-

FIG. 50.—PLASTER-JOINTS.

tageous when making repairs, and is made by spiking and bolting a piece of plank 3 in. thick, 12 in. wide, and 3 ft. long to each side of the cap or sill, as the case may be, and to each post.   This joint has been adopted by the Delaware and Hudson Canal Co., and is said to be proving very satisfactory.   The details are shown in Fig. 50.   With this joint all the posts should be notched 1 in. to both sill and cap.

There is a joint in use on the New York, Lake Erie & Western Railroad,* made with an iron plate bent in a special manner, and which allows of the very easy removal of parts for repairs, while at the same time it is strong and efficient.   Fig. 51 shows this joint in all its details.†

Nearly every conceivable combination of the above joints with or without notching is in use.   For batter-posts, the notch shown in Fig. 52 is rather better than that in Fig. 53.

The height of the bent is measured from the under side of the sill to the top of the cap.

FIG. 51.—IRON JOINT-PLATE, N.Y., L. E. & W. R. R.

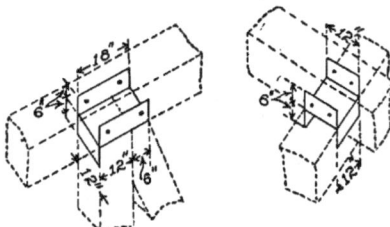

FIG. 52.   FIG. 53.
BATTER-POST NOTCHES.

not be quoted as its height, as is frequently, though wrongly, done.

Bents should be spaced at such a distance between centres as will use the length of timber easiest to obtain for stringers in the most economical manner.   The distance varies from 12 ft. to 16 ft.; spans of 14 ft. and 15 ft. being the most general.   Where it is possible, all the bents should be evenly spaced, only employing spans of unequal length where they cannot be avoided.

That which was said in the chapter on Pile-bents in relation to split-caps applies with the same force to framed bents.

Both the sills and caps on the Savannah, Florida & Western Railroad, W. B. W. Howe, Jr., Chief Engineer, are split horizontally, the upper and lower pieces being held together, and kept from sliding by pins driven into holes bored through them.

---

* Also used on the Chicago, Rock Island & Pacific Railroad.
† *Engineering News,* Nov. 5, 1887.

CHAPTER V

FLOOR SYSTEM.

Corbels.—Corbels are pieces of timber placed lengthwise of the stringers, between them and the caps. They are usually from 4 ft. to 8 ft. long, extending equal distances on either side of the centre of the cap. They are not much in favor, for good reasons. To a certain

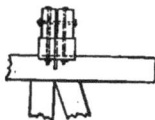

FIG. 54.—DELAWARE & HUDSON CANAL CO.

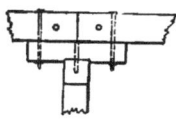

FIG. 55.—CHARLESTON, CINCINNATI & CHICAGO R. R.

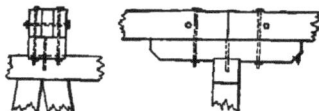

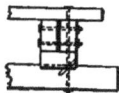

FIG. 56.—LOUISVILLE & NASHVILLE R. R.

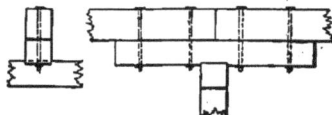

FIG. 57.—SCIOTO VALLEY R. R.

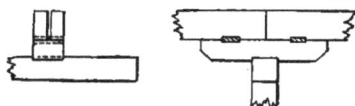

FIG. 58.—OHIO CONNECTING R. R.

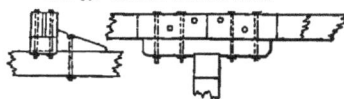

FIG. 59.—NEW YORK, LAKE ERIE & WESTERN R. R.

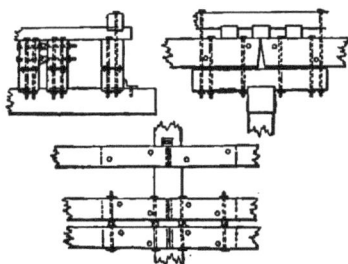

FIG. 60.—CHICAGO & NORTHWESTERN R. R.

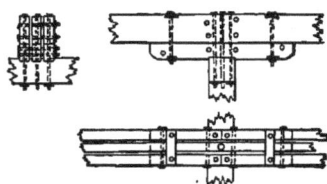

FIG. 61.—NEW YORK, WOODHAVEN & ROCKAWAY R. R.

FIGS 54 TO 61.—DETAILS OF CORBELS.

extent they are very useful, but they also have many disadvantages. They give extra support to and consequently strengthen the stringers; but for various reasons, as the stringers should not be made lighter on this account, this does not count for much. They also add stiffness

31

to the stringer-joint, but sufficient stiffness for all intents and purposes may be obtained from a well-designed joint without them. They add to the cost, not only in labor and lumber, but also require the use of a considerably larger amount of iron. They increase the number of joints, and hence the places for the lodgment and beginning of decay. If, however, it is thought desirable to use them, the different ways of fastening the stringers to them, and they in turn to the caps, may be seen in Figs. 54 to 61.

Corbels should be notched down about 1 in. over the cap. A peculiar and rather commendable method of separating the corbels and stringers from each other by cast-iron blocks, as adopted on the Chicago & Northwestern Railroad, is shown in Fig. 60.

Stringers.—A stringer should be placed immediately beneath each rail, and in order to guard against defective timber it ought to be "split" or composed of two or more pieces. These pieces should be separated from each other by either cast-iron washers or spools, or wooden packing-blocks, or both. A considerable difference exists in the present practice as to the amount of separation. It varies all the way from nothing to 13 in. From $1\frac{1}{2}$ in. to 2 in. is a very good distance. In Figs. 62 to 72 are shown a number of cast-iron separators, and in Figs. 73 to 82 a number of wooden packing-blocks. Among the latter, those having the general form of Fig. 75 are to be preferred. These are to be placed immediately above the caps. Those packing-blocks which are notched are of course placed so that the cap fits

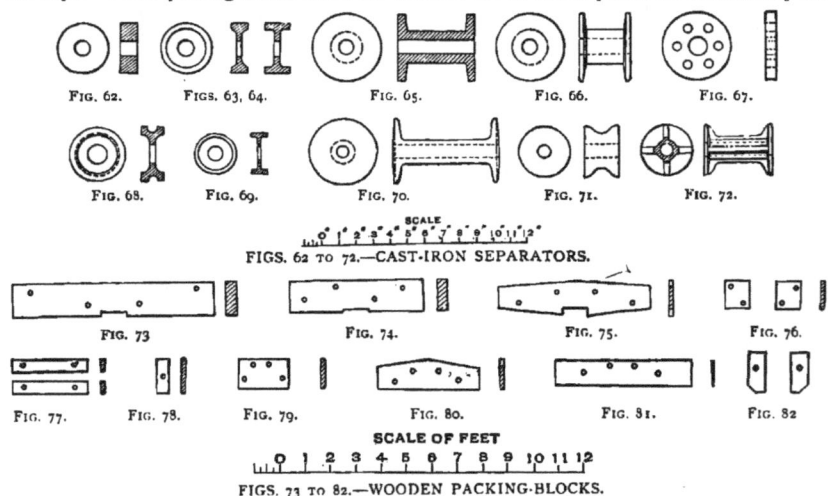

FIG. 62.   FIGS. 63, 64.   FIG. 65.   FIG. 66.   FIG. 67.

FIG. 68.   FIG. 69.   FIG. 70.   FIG. 71.   FIG. 72.

SCALE

FIGS. 62 TO 72.—CAST-IRON SEPARATORS.

FIG. 73   FIG. 74.   FIG. 75.   FIG. 76.

FIG. 77.   FIG. 78.   FIG. 79.   FIG. 80.   FIG. 81.   FIG. 82.

SCALE OF FEET

0 1 2 3 4 5 6 7 8 9 10 11 12

FIGS. 73 TO 82.—WOODEN PACKING-BLOCKS.

into the notch. Frequently the packing-blocks made of the heavier material are used merely as splice-blocks, they being separated from the stringers by thin cast-iron separators, such as is shown in Fig. 67. Many fasten the stringers together by intermediate bolts placed either at the centre of the span or at regular intervals along it. Separators or packing-blocks are of course required to be placed between the stringers wherever these bolts are located. With good timber and spans of 12 ft. to 14 ft. these intermediate bolts are not necessary, and may be just as well omitted as not.

When it is possible, the stringer-pieces should be long enough to extend over two spans and the joints broken. Various styles of stringer-joints and ways of arranging intermediate bolts are shown in Figs. 83 to 98. The arrangement shown in Fig. 83 is to be greatly pre-

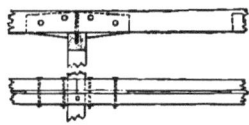

Fig. 83.—Pennsylvania R. R.   Fig. 84.—Wisconsin Central R. R.   Fig. 85.—N. Y., P. & B. R. R.

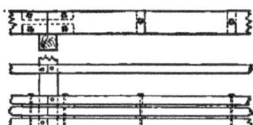

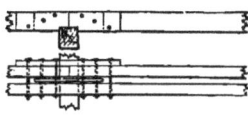

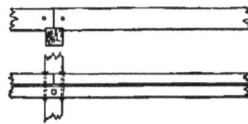

Fig. 86.—B. & M. R. R. in Neb.   Fig. 87.—A. & P. R. R.   Fig. 88.—Georgia Pacific Ry.

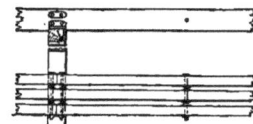

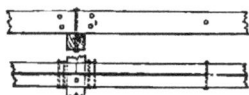

Fig. 89.—C., N. O. & T. P. Ry.   Fig. 90.—Central R. R. of Ga.   Fig. 91.—Gulf, Col. & Santa Fe R.R

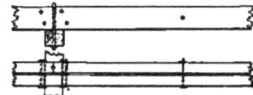

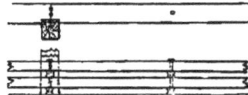

Fig. 92.—D., T. & Ft. Worth R. R.   Fig. 93.—Chicago & W. Mich. Ry.   Fig. 94.—Chicago & Atlantic Ry.

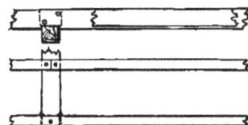

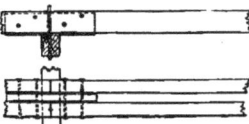

Fig. 95.—B., C. R. & Northern R.R.   Fig. 96.—San F. & N. Pacific R. R.   Fig. 97.—St. P., Min & M. Ry.

Fig. 98.—Oregon Pacific R. R.

FIGS. 83 TO 98.—DETAILS OF STRINGER-JOINTS.

ferred, because, should the support for any reason become weakened, the joint, when it settles as a weight comes upon it, closes at the top and tends to open at the bottom. Now the lower bolts act somewhat as a fulcrum, and the effect will be to tend toward splitting the stringer from these bolts to the nearest end. As this arrangement gives the most material where there is the greatest liability to split, and consequently at the weakest point, it forms the strongest kind of a joint.

Such joints as those shown in Figs. 91, 97, and 98 cannot be condemned too strongly, and are always to be avoided. Those illustrated in Figs. 87, 88, 89, 90, 92, 93, 94, and 96 are also poor on account of the packing-bolts being so close to the end of broken stringer-pieces, and also, in some cases, on account of there being too few of them. That in Fig. 92 would be an excellent joint were the lower bolts placed a foot or so farther apart. The joint shown in Fig. 86 is said by Mr. I. S. P. Weeks, Chief Engineer C., B. & Q. R. R. west of the Missouri River, to have proved very efficient. It has carried an engine over after the bent has been washed out.

The bolts holding the stringer-pieces together, and which are called packing-bolts, should be long enough to extend clear through from face to face of the complete stringer, and allow of placing a cast-iron washer under both nut and head.

When the stringers are not fastened directly to the caps they should be notched over

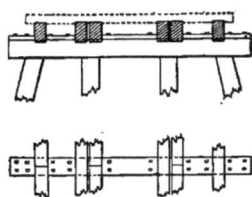

FIG. 99.—STRINGER FASTENING.

them 1 in. A method for holding the stringers in place, and which is becoming quite general, is shown in Fig. 99. It consists of a piece of 3 in. × 12 in. plank, fastened, outside of each stringer, to the cap by four log-screws or by spikes. The stringers in their turn are kept at the proper distance apart either by a spreader made of the same material or by fastening the ties to them.

The size of the stringer-pieces in cross-section will vary with the span, variety of timber, and weight of the traffic. They should be of sufficient dimensions to prevent any considerable deflection by a passing train. For long spans, or on lines having heavy loads and engines, each stringer should be composed of three pieces; in other cases two are sufficient. The practice of the Pennsylvania Railroad in this respect is given in Table IX.

TABLE IX.

Trestle-stringers, Pennsylvania Railroad Standard.

Dimensions of Stringers.			
Clear Span.	Number of Pieces under each Rail.	Width of each Piece.	Depth of Stringers.
10 ft.	2	8 in.	15 in.
12 "	2	8 "	16 "
14 "	2	10 "	17 "
16 "	3	8 "	17 "

A "jack-stringer," composed of a single piece, should always be placed under either end of the ties, as in Fig. 99. By such an arrangement many advantages are secured. The principal one is in case of a derailment, when, if the ties give way, the cars are not liable to fall to the ground as they otherwise might. As the ends of the ties are supported, the chances are very much in favor of their not being broken in such a case. Thus the factor of safety is largely increased. These outer stringers should be long enough to extend over two spans, and should always be securely fastened to the caps by a drift-bolt through either end and the centre.

The ends of the stringer-pieces are generally butted together. There are two exceptions to this otherwise universal rule: in the trestles on the San Francisco & North Pacific Railway, Fig. 96, the ends are separated ¾ in., and in those of the Chicago & Northwestern Railroad, Fig. 60, they are bevelled 1 in.

Several roads have adopted the policy of trussing stringers having a span of 14 ft. or over after they become three or four years old. This end is accomplished on the Pontiac, Oxford & Port Austin Railroad, Geo. A. Nettleton, Chief Engineer, by arranging an iron rod and pieces of rail as shown in Fig. 100. While this treatment has a very beneficial effect in some respects, and adds considerably to the strength of the structure, still it seems as though the men in charge of the trestles, as well as the inspectors, would be

FIG. 100.—TRUSSING STRINGERS.

tempted to rely too much upon this extra strength, and allow timber to remain in service which should for safety have been removed long before. The carelessness which would thus tend to be inculcated, would prove very dangerous on the majority of roads.

Ties.—Ties may be of 6-in. × 8-in. timber, sawed, and should have a length of 12 ft. They should be notched over the stringers 1 in., and if outside stringers are used with notched guard-rails they need not be otherwise fastened. In other cases they should be spiked to the stringers. There are many different ways of arranging the spikes. Some fasten every third or fourth tie only, while others spike every tie. It is always better to stagger the spikes or arrange them zigzag, as in Fig. 101. Figs. 101 to 109 show several of the different ways of arranging these fastenings. Opinions as to the spacing of the ties vary. They are placed anywhere from 12 in. to 24 in. from centre to centre. The closer together they are put the better; they should never be spaced with centres over 12 in. apart, leaving 6-in. openings between the ties; 9-in. centres are far better even than 12-in. On the West Shore Railroad small blocks 4 in. thick × 8 in. square are spiked to the stringers between the ties in such a manner as to act as a cover for the space between the stringer-pieces (Fig. 101). While these blocks serve a good end by preventing "bunching," and in keeping out the rain and moisture, they are hardly advisable because of their interfering with the free circulation of the air between the separate pieces of the stringer, as well as on account of their preventing the penetration of the sunlight into these places.

Often when the ties are not notched, and it is desired to use some other form of fastening than spiking, dowel-pins, made of ¾-in. iron 5 in. long, may be resorted to. They may be arranged as in Fig. 109.

Guard-rails.—Guard-rails serve two principal purposes: first, to keep the train from leaving the bridge in case of a derailment; and second, to aid in keeping the ties in their proper places, and give stiffness to the floor system. They should always be employed, and where an outside stringer is used should be placed immediately above it. They need not be made of very heavy timber, nor should they be too light; 6 in. × 8 in., with the narrow face down, is a very good size. The length may vary, using such timber as can be most conveniently obtained; still it is better to have them from 16 ft. to 20 ft. long. Of course greater length is in no wise objectionable, except that it is rather more difficult to obtain, and hence

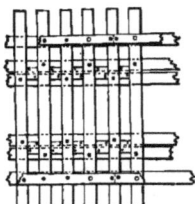

FIG. 101.—N. Y., W. S. & B. R. R.

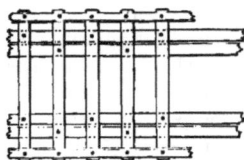

FIG. 102.—PENNSYLVANIA R. R.

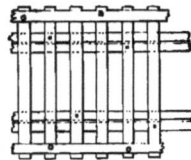

FIG. 103.—TEXAS & PACIFIC RY.

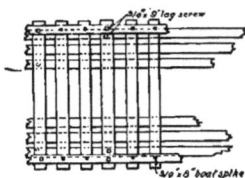

FIG. 104.—T., ST. L. & K. C. R. R.

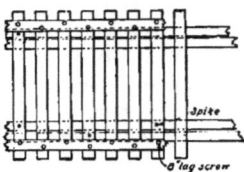

FIG. 105.—K. C., F. S. & M. R. R.

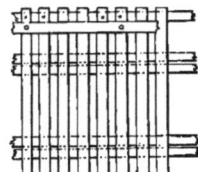

FIG. 106.—ST. P., M. & M. RY.

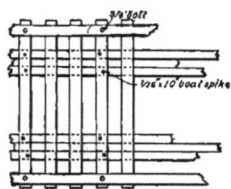

FIG. 107.—C., C. & C. R. R.

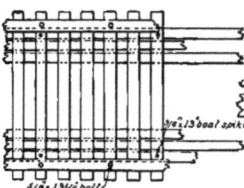

FIG. 108.—M., K. & T. RY.

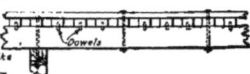

FIG. 109.—L. & N. R. R.

FIGS. 101 TO 109.—FLOOR SYSTEMS.

more costly. There are a number of forms of joints in use for connecting the pieces together. Many of these are shown in Figs. 110 to 115. The ordinary halved joint, Fig. 114, is an excellent one, and fully answers all requirements. The joints should always come immediately over a tie and be broken; i.e., those on opposite sides should be over different ties, no two joints coming over the same tie. A bolt should extend through the joint tie and outside stringer. The guard-rail should always be notched down at least 1 in. over each tie.

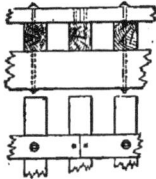

FIG. 110.—N. Y., W. S. & B. R. R.　　FIG. 111.—PENN. R. R.　　FIG. 112.—T., ST. L. & K. C. R. R.

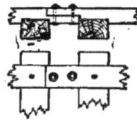

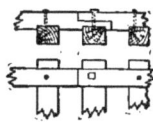

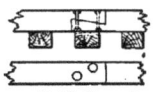

FIG. 113.—C., M. & ST. P. RY.　　FIG. 114.—R. & D. R. R.　　FIG. 115.—N. Y. ELEV. ROADS.

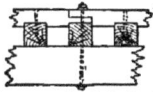

FIGS. 110 TO 115.—GUARD-RAIL JOINTS.

The ends of the guard-rails at either end of the bridge ought to be rounded off or cut at an incline, as in Figs. 116 and 117. Every tie should be fastened to the guard-rail in some way, especially when they are not fastened to the stringers. A bolt should be put through the guard-rail at every fourth or fifth tie, and should extend through the outside stringer. The balance of the ties may be spiked or fastened by lag-screws. Spiking is much cheaper, a $\frac{1}{2}$-in. $\times$ 10-in. boat-spike being employed. If lag-screws are used, a $\frac{5}{8}$-in. $\times$ 8-in. screw is a very good size. A wrought washer is to be placed under the head of each lag-screw, and a 3-in. to $3\frac{1}{2}$-in. cast washer under the head and nut of each bolt. The screw or nut ends of the bolts should be placed up so that they may be more easily inspected and tightened. It is not necessary to countersink the nuts of the bolts or the heads of the lag-screws; in fact it should not be done unless absolutely unavoidable, as the holes form a basis for the lodgment of water, and thus are apt to prove very harmful. At either end of the bridge the guard-rails should extend at least from 20 ft. to 30 ft. on to the embankment, and be flared to such an extent that their extreme ends will be the gauge of the track from the rails. They should be supplemented by bumping-posts (Fig. 118). These, however, will be spoken of later on. It is better, though of course more costly, to face the inside upper corner of the guard-rails with angle-iron. This overcomes to a very large extent the tendency of the wheels to override the guards, by preventing the wheels from cutting into them. Frequently the upper edges of the guard-rails are bevelled. This is bad practice, as it reduces the effective height of the guard, and tends to assist the wheels in overriding them.

Inside guard-rails, either of wood or of a second steel rail, placed about $2\frac{1}{2}$ in. from the rails, are claimed by many to be much more efficient than outside guards. Outside guards, it is said, tend to turn a derailed truck at right angles to the moving train, while inside guards turn it towards the track. It is urged against inside guards that articles such as brake-shoes, etc., are very apt to fall between the guard and the rail, and thus increase the number of derailments. However this may be, there is no doubt that inside guards are very serviceable, but their use is no reason for omitting the use of outside guards, which should always be employed. In regions where it is necessary to use snow ploughs on roads where the pilot comes very close to the rails, inside guards should never rise above the top of the rails.

FIG. 116.　FIG. 117.
GUARD-RAIL ENDS.

Fastening down Floor System.—There are a number of different methods of fastening the floor system to the bents, some of which have already been described.   Drift-bolting the stringers to the caps is the one most generally employed.   The drift-bolts should extend a

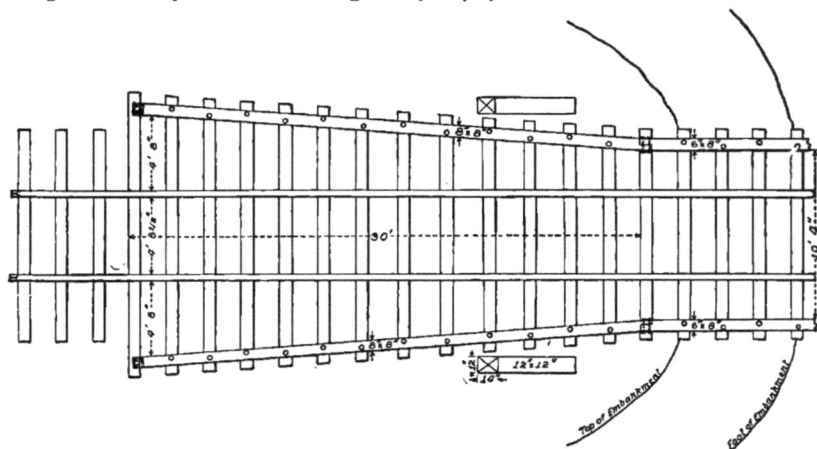

FIG. 118.—EMBANKMENT END OF TRESTLE, SHOWING FLARED GUARD-RAILS AND BUMPING-POSTS.

generous distance into the caps,—say at least 8 in.   One drift-bolt through the continuous piece of each compound stringer, per bent, especially if the ties are notched, is amply sufficient (Fig. 119).

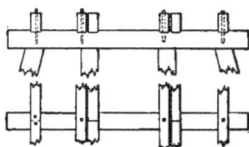

FIG. 119.—DRIFT-BOLTING-DOWN STRINGERS.

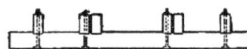

FIG. 120.—BOLTING-DOWN STRINGERS.

Among the other ways is that of using a ¾-in. bolt with nut in place of a drift-bolt (Fig. 120).   This bolt is sometimes made long enough to extend through a tie placed immediately above the cap, in which case it usually passes through the space between the stringer-pieces

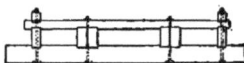

FIG. 121.

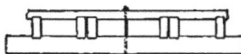

FIG. 122.

BOLTING-DOWN STRINGERS.

(Fig. 121).   Several roads employ but one bolt, placed on the centre line, as in Fig. 121.   Frequently the floor system is not fastened to the bents at all, its weight being depended on to keep it down, and blocks arranged as shown in Fig. 99, and Plates II, III, XXIV, XXVII, etc., Part II, to keep it in place and line.   In this case girts, securely fastened to the posts at their upper ends, should always form a part of the structure, no matter how low it may happen to be.

CHAPTER VI

BRACING, COMPOUND-TIMBER TRESTLES, HIGH TRESTLES, TRESTLES ON CURVES, AND MISCELLANEOUS TRESTLES.

Sway-bracing.—It is seldom that any sway-bracing will be needed for either pile or framed bents under 10 ft. high. For those from 10 ft. to 20 ft. in height a single X of 3-in. × 10-in. plank is all that is necessary. One plank should be placed on either side of the bent, and extend from the upper corner of the cap across to the lower end of the opposite batter-pile, terminating just above the ground, or to the opposite lower corner of the sill if a framed bent. The braces should be bolted to the cap, to each pile or post, and to the sill by a ¾-in. bolt, with a cast washer under both head and nut. Often either lag-screws or spikes are used for attaching the braces, but bolts are to be preferred.

For bents over 20 ft. high but not over 40 ft. two X's of sway-bracing should be employed. It is both more convenient and more economical to make the upper X of a constant length, say from 15 to 20 ft., and put the odd lengths in lower one. A horizontal stick on each side of the bent separates the X's. These sticks are also made of 3-in. × 10-in. plank, and bolted to each post or pile.

Whenever a pile or a cap extends beyond the other so that the sway-braces cannot lie flat, either the larger of the two should be sized down so as to be level with the smaller, or else the smaller should be blocked out to meet the brace. In general, the former method is the better one.

Counter-posts.—When framed bents approach a height of 40 ft., they are frequently stiffened by the use of counter-posts rather than sway-bracing, though sometimes by the use of both. The employment of counter-posts requires the dividing of the bent into two stories by means of an intermediate sill. Plates XV, XIX, XXI, XXIII, XXIV, Part II, show several methods of using counters. They are more generally employed in very high work, and for further particulars in regard to them the reader is referred to the section on High Trestles.

Longitudinal Bracing.—There is considerable variation in the methods of longitudinal bracing employed, some bracing every bay, others only every third or fourth; some arranging the braces diagonally or latticed, others horizontally, and still others in what might be called a laced form. Examples of all of these forms are shown in Plates XVI, XXIX, XI, Part II. All possible combinations of these, especially of the last two, are employed, as well as many modifications and adaptations. Plate XI illustrates that which may be called the laced form, and is the standard on the Pennsylvania Railroad. The ends of the braces are cut in the form, and the edges of the caps and sills chamfered, as in the detail drawing. Each piece is fastened to both cap and sill by a heavy cut spike. There is but one stick of 8-in. × 8-in. material to each bay, and it is placed in the centre line of the trestle. When horizontal bracing, such as shown in the side elevation in Plate XXIX, Part II, is used, there should be a stick placed

39

immediately above the sill on the outside of each post, and one immediately above the horizontal piece of the sway-bracing.

Lateral Bracing.—Lateral bracing, such as is illustrated in Plate IX, Part II, adds very greatly to the stiffness of a structure.* It is made of two 6-in. × 6-in. timbers placed diagonally across, from cap to cap, immediately beneath the stringers and bolted together at the intersection by a ⅝-in. or ¾-in. bolt. The timbers are usually slightly notched into the caps, and fastened in place by several heavy spikes. This kind of bracing is coming into quite general use, and is now one of the essentials of many new designs. When used, the longitudinal bracing need not be so extensive. It is said that where lateral bracing is employed the trestle keeps in line much better.

Compound-timber Trestles.—There is a style of construction very largely in vogue which may be denominated as above. The members, such as caps, sills, posts, etc., either wholly or partly, are each composed of two or more pieces bolted together instead of being a solid stick. The parts are generally separated from each other to a greater or lesser degree. While the life of the structure may be somewhat shortened in some cases, it is claimed that this disadvantage is more than offset by the ease of repairs, as any part can be replaced with a minimum amount of labor, and without causing the least disturbance in the running of trains or impairing the safety in any way. On account of the smaller size of the timber, much more thoroughly seasoned and better quality material can be obtained. It can also be much more easily inspected. The sticks are generally 6 in. × 12 in. Several plans of this style of structure are given in Plates XXVII, XXVIII, XXIX, XXX, and XXXII, Part II.

High Trestles.—Trestles above 40 ft. in height may be classed as high trestles. Usually they are divided into two or more decks and stories. The height of the decks depends upon several considerations, but is regulated to a certain extent by the length of timber that can be most economically procured. The decks and stories should be of uniform height throughout any one trestle, or at least those upon the same level should be, in order to simplify things as much as possible, and the odd lengths put into the lowest one. It is in the designing of these rather exceptional structures, especially when the extraordinary height of one hundred or more feet is reached, that there is every opportunity for the full employment of a very high grade of constructive skill.

There may be said to be four classes of high trestles :

1st. Those in which the posts are continuous, being made up to the required length by joining single sticks together, end to end, with a butted joint, using splice-blocks or other means.

2d. Those in which the decks, though separate and distinct, are still intimately joined together by means of framing ; the sill of one deck acting as the cap of the one beneath.

3d. Those in which the decks are separated entirely by purlins or other means.

4th. Those in which the posts, and frequently other members, are each made up of two or more pieces placed together side by side. In this latter group are included cluster-bent trestles.

---

* Adopted as standard on the Boston & Albany Railroad; the Toledo & Ohio Central Railroad, C. Buxton, Chief Engineer.

Those of the first class are generally erected where good quality long timber of large size may be easily and economically procured. In this group stories are formed by bolting horizontal pieces of timber to the posts, one on either side, at the proper heights. Counter-posts, or what may be called inside batter-posts, are often introduced, a new set being put in at every other story, and continued down to the main sill. This class of trestle is shown very clearly in Part II, Plates XV, XVI, and XVIII. In Plates XV and XVIII the employment of counters is depicted.

Classes 2 and 3 are resorted to when but comparatively short timber can be procured, and for several reasons are, in the writer's judgment, rather to be preferred, especially the third class, to the continuous-post group. The second class hardly needs any enlargement, as the mere defining of it at once describes its peculiarity. The posts are generally connected with sills and caps in this type by mortise and tenon joints. All posts should of course come immediately beneath those in the deck above, and be in the same line with them, forming to all intents and purposes a continuation of them. Illustrations of this type are given in Part II, Plates XIX to XXII.

In the third class the bents of each deck are distinctly separate, being framed entirely by themselves. The lower-deck bents are erected, and then purlins laid along on the caps in such a way as to come directly under the posts of the deck above, the bents of which are of course placed directly over those of the one below. Purlins are laid on the caps of these, and the next deck erected on top of them. This is continued until the necessary height has been attained. The purlins should be firmly fastened to the caps on which they rest either by ordinary bolts or by drift-bolts. The sills should also be secured to the purlins underneath them in the same manner. For illustrations of this construction see Plates XXIII to XXVI. This style offers many advantages for ease of erection, which will be more readily appreciated when that subject is treated of.

The fourth class may be subdivided into two groups, namely, those in which the posts, and sometimes other members, are built up by bolting two or more pieces together, keeping them separated a little from each other (see Part II, Plates XXVII, XXX, XXXI, etc.),—the majority of them might almost be called plank trestles,—and those in which each post is made up of four smaller posts, two of the smaller posts always being continuous over any one story: these are known as cluster-bent trestles. Both of these styles are claimed to have a number of advantages over those built with single sticks of large dimensions. Among them may be mentioned the ability to secure better material, both as respects quality and seasoning, on account of the pieces being smaller; greater economy and ease in the cost of erection; and especially greater facility for making repairs. It is also claimed that they can be much more thoroughly, easily, and certainly inspected. While it is said that their life is hardly as long as that of the others, still the advantages enumerated, it has been stated, greatly offset this disadvantage. Besides, they may be kept in a much safer condition.

In all of the different styles the bents should always be thoroughly sway-braced, each story and deck having its own set of braces. There should always be, also, a set of longitudinal braces to each deck. As a rule these are of the horizontal type. It should not be attempted to economize in the amount of timber by reducing either the number or the size of the girts.

Scanting the amount of longitudinal bracing is in no case real economy: it is in fact outrageous, tending to great danger to human life. Frequently two adjacent bents every three or four bents apart are connected by diagonal longitudinal braces so as to form, in effect, towers similar to those of iron trestles. While this is an excellent plan, the longitudinal bracing of the intermediate spans should not be left out, as is generally the case; for while the tower construction adds considerably to the stiffness of the structure as a whole, it is no excuse whatever for weakening the remaining parts. To the writer it seems that the best form of high trestle is the cluster-bent type, with every third bay braced diagonally so as to form a tower, and with the intermediate bays braced with horizontal sticks at every deck, a 3 × 10 in. plank being placed on each side of every post.

The plentiful use of counter-posts is also to be recommended. For giving lateral stiffness to the structure, the lateral bracing described on page 40, and illustrated in Part II, Plate IX, is very effective, and should be used whenever possible.

Considerable economy may be effected in trestles of great height by spacing the bents farther apart, say thirty feet, and supporting the floor on a deck truss. Such a construction is shown in Part II, Plates XXXIII and XXXIV.

The floor system for high trestles is of course the same as that for the lower structures, and which was discussed fully in Chapter V.

A far more thorough knowledge of the various practice in the treatment of these structures may be obtained by the careful study of the plates in Part II, than could be imparted by mere descriptive matter, and so the reader is referred to them.

Trestles on Curves.—Of course, whenever it is possible, building a trestle on a curve should be avoided. Sometimes, however, this cannot be helped, and then we have to resort to the best means at our command to increase their strength and safety. It is preferable to place the bents on radial lines, especially where the curve is a sharp one. The bracing of all kinds should be heavier and more abundant than where the structure is on a tangent. It is also well to give the batter-posts, especially those on the outside of the curve, as much inclination as possible, a batter of $3\frac{1}{2}$ in. to 4 in. per foot not being out of the way, so as to increase the breadth of the base, and enable the trestle to better resist the centrifugal force of the train.

Lateral bracing should always be employed on curved trestles, as it tends to save the structure considerably from the racking it otherwise receives from the train.

There are six different ways of elevating the outer rail in common use. The first is by cutting the piles or posts so as to give the cap the proper inclination; the second, by tapering the tie, as in Plate XXXVI, Part II; the third, by placing wedge-shaped blocks between the tie and the stringer, and bolting them to the former, as in Fig. 123; the fourth, by shiming up the track by wedge-shaped blocks placed on top of the ties, and securely spiked to them, as in Fig. 124; the fifth, by placing a bolster or corbel under the stringers on one side and not on the other; and the sixth, by notching or sizing down one end of the cap, as in Fig. 125. This latter method is in use on the Clinch Valley Division of the Norfolk & Western Railroad, and the dimensions given in the figure are for a 6° curve. The other methods are in use on numerous roads throughout the country.

One very serious objection urged against the second method is the splitting of the tie by the jarring of the trains; another, the increased cost of the tie, because a tie having the cross-section of the largest end has to be paid for. When this method is used, the face on which the fibres are cut across diagonally should always be placed down.

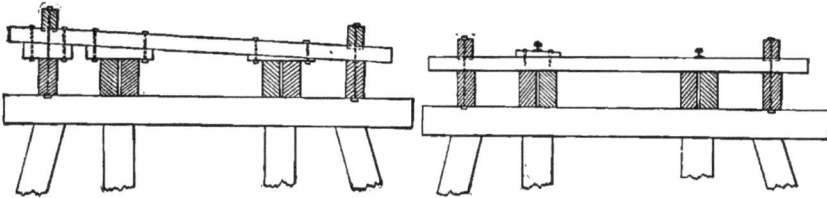

FIG. 123.   FIG. 124.

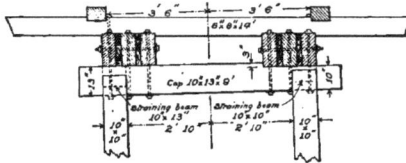

FIG. 125.

FIGS. 123 TO 125.—METHODS OF ELEVATING TRACK ON CURVED TRESTLES.

Examples of trestles built on curves are given in Part II, Plates IX and XXXVI.

Double-track Trestles.—Double-track trestles, as a rule, are little else than two single track trestles placed side by side and intimately joined together. The caps and sills should always be continuous. The two batter-posts or piles which would come in the centre are replaced by a single vertical post or pile, or else entirely omitted, and a heavy guard-rail is bolted to the ties about half-way between the inside rails. An outside or "jack" stringer should always be placed beneath this guard-rail, and secured firmly in place. No scanting of the fastening on account of its interior position should be allowed. Plates IX and XXXV to XXXVII, Part II, show several double track trestles.

Knee-braced Trestles.—On unimportant branch lines, where the traffic is light and the trestles high, considerable economy in timber is attained by using the knee-braced type of trestle. In this form every other bent is omitted, making the spans just twice the ordinary length. The stringers are strengthened by placing a short straining-beam beneath them, and running knee-braces from either end of it down against the posts. Many engineers object very strongly indeed to using this form of construction at all. Plate XVI, Part II, shows a form of this type of trestle, which is the standard on the Norfolk & Western Railroad.

Round-timber Trestles.—It frequently happens that it is rather difficult to obtain sawed timber, and extensive hewing is both expensive and unnecessary. In this case the trestle is built of round timber. This form of structure is exceedingly cheap, and if well built is very serviceable, though rather rough and unfinished in appearance.

Trestles with Solid Floors.—On the line of the Louisville & Nashville Railroad, between Mobile and New Orleans, there are some trestles of very peculiar construction. The floor is made in the form of a trough and filled in with earth. The ties and rails are then laid on top of this filling, the same as on an ordinary embankment. For certain climates and regions this construction has much to recommend it. It is especially adapted to mild southern climates, and is almost absolutely protected against destruction by fire from cinders dropped by a locomotive. All of the timber should be thoroughly creosoted Plate VIII, Part II, shows, very clearly, one of these trestles.

CHAPTER VII.

IRON DETAILS.

Spikes.—There are two varieties of spikes used in trestle-building,—cut spikes and boat or ship spikes. Cut spikes (Fig. 126) are fashioned after the same pattern as common nails, and are essentially stamped out of sheet-metal. They should be of good quality and have generous-sized heads. Table X gives the number of cut spikes in a keg of 100 lbs., and also the weight in pounds of a single spike.

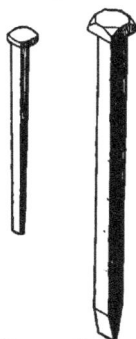

TABLE X.

Cut Spikes.

Length in inches.	No. in Keg, 100 lbs.	Weight of one Spike, lbs.	Length in inches.	No. in Keg, 100 lbs.	Weight of one Spike, lbs.
3	2900	.0344	5½	850	.1176
3½	2100	.0476	6	775	.1293
4	1500	.0667	6½	575	.1739
4½	1150	.0869	7	450	.2222
5	950	.1052	8	375	.2666

FIG. 126. FIG. 127.
CUT BOAT-
SPIKE. SPIKE.

Occasionally common nails of the larger sizes have a limited use, and as an aid in estimating, Table XI, giving their size and weight, is appended.

TABLE XI.

Size and Weight of Nails.

Name.	Length.	No. in a lb.
10-penny common.	3 inches	60
12 " "	3¼ "	44
16 " "	3½ "	32
20 " "	4 "	24
30 " "	4½ "	18
40 " "	5 "	14
50 " "	5½ "	12
60 " "	6 "	10
8 " fence.	2½ "	50
10 " "	3 "	34
12 " "	3½ "	29

These nails are of the same pattern as the spike shown in Fig. 126, but smaller. Boat-spikes are forged from bars of wrought-iron, and are of the general shape shown in Fig. 127. They have a square section, and are sharpened at the end to a kind of blunt chisel-point. This kind of spike is the one most commonly used in building trestles, and is always the kind to be employed in fastening guard-rails to ties and ties to stringers. Table XII gives the

45

approximate number of boat-spikes in a keg of 150 lbs. in heavy-faced type, and the weight of a single spike in light-faced type.

TABLE XII.

Number of Boat-spikes in a Keg of 150 lbs. and Weight of a Single Spike.

Thickness Ins.	Length in Inches.														
	3	3½	4	4½	5	5½	6	6½	7	7½	8	8½	9	9½	10
¼	1910 .0785	1585 .0946	1326 .1093	1223 .1226	1025 .1463										
⁵⁄₁₆	1010 .1485	963 .1557	810 .1851	605 .2479	583 .2572		521 .2879								
⁷⁄₁₆			542 .2767	503 .2982	461 .3253	423 .3546	402 .3731	321 .4673							
½					340 .4117	312 .4839	298 .5033	280 .5357	261 .5747	240 .625	223 .6726				
⁹⁄₁₆							221 .6787	200 .75	190 .7881	180 .8333	170 .8823	160 .9375	150 1.0000	140 1.0714	130 1.1538
⅝											140 1.0714	130 1.1538	120 1.25	110 1.3636	100 1.5000

Drift-bolts.—The common form of drift-bolt is but little else than a very long boat-spike, though other shapes are used quite extensively. They should always be long enough to penetrate the last timber desired to be held to a depth sufficient to give a good firm hold.

Fig. 128 gives the forms of bolts in general use, the first one being that most commonly employed. They are usually made of iron having a section ¾ in. square or a diameter of ¾ in., and for fastening 12-in. caps to posts or piles are generally 20 in. long. Their weight is about as given in Table XIII.

FIG. 128.—DRIFT-BOLTS.

TABLE XIII.

Weight of Drift-bolts.

Length in Inches.	Square Section.		Round Section.	
	¾" Sq.	1" Sq.	¾" Diam.	1" Diam.
	lbs.	lbs.	lbs.	lbs.
18	2.9	5.1	2.3	4.0
20	3.2	5.7	2.5	4.4
22	3.5	6.2	2.8	4.9
24	3.8	6.8	3.0	5.3
26	4.1	7.3	3.3	5.8

The main value of drift-bolts lies in their holding power. Following is a summary of three series of experiments upon this subject:*

---

* *Engineering News*, Feb. 28, 1891.

U. S. Government Experiments.—These experiments were made under the direction of General Weitzal by Assistant U. S. Engineers A. Noble and C. P. Gilbert, in 1874–77, and were published by Colonel O. M. Poe in his report to the Chief of Engineers for 1884. This series was very extensive, but the valuable results obtained are robbed of much of their value by the lack (in the original publication) of suitable comparisons and conclusions.

The mean of from 150 to 200 experiments with round and square bolts, both ragged and smooth, in different-sized holes, shows that the resistance after having been driven seven months is 10 per cent greater than the resistance immediately after driving, the different sizes and forms being strikingly uniform. The mean of 150 experiments under various conditions shows that the resistance to being drawn in the direction which it was driven is only 60 per cent of its resistance to being drawn in the opposite direction; that is to say, the resistance to being drawn *through* is only 60 per cent of that to being drawn *back*. The mean of 50 experiments shows that smooth rods have a greater holding power, both to being drawn through, and also to being drawn back, than ragged ones, a "moderate ragging" reducing the resistance a little more than 25 per cent, and an "excessive ragging" reducing the holding power more than 50 per cent.

Concerning the best relation between the diameter of the bolt and that of the hole, one series of 60 experiments, shows that the holding power of a 1-in round rod in a $1\frac{1}{8}$ hole is greater than in either a $1\frac{3}{16}$ or in a $1\frac{3}{8}$ hole, the resistance in the $1\frac{3}{16}$ hole being 98 per cent, in the $1\frac{3}{8}$ 90 per cent, of that in the $1\frac{1}{8}$ hole. On the other hand, another series of 35 experiments makes the resistance in a $1\frac{2}{16}$ hole greater than in a $1\frac{3}{16}$ or a $1\frac{1}{4}$, the first two being practically the same, and the last being only 85 per cent of the first. However, the difference between the two series is not material, considering the nature of the experiments. For a $\frac{3}{4}$-in. round bolt, four experiments on each size seem to prove that the holding power in a $\frac{13}{16}$ hole is about one quarter greater than in a $\frac{9}{16}$ or an $\frac{11}{16}$ hole. For a 1-in. square bolt, the holding power in a $1\frac{1}{4}$ hole is only a trifle greater than in a $1\frac{3}{8}$, and about 20 per cent greater than in a $1\frac{13}{16}$ hole, as deduced from 20 to 40 experiments for each size of hole.

The holding power of a 1-in. square bolt in a $1\frac{1}{4}$ hole was practically the same as for a 1-in. round rod in an $1\frac{1}{8}$-in. hole. There is 25 per cent more metal in the square drift-bolt, while more labor is required to bore a $1\frac{1}{4}$-in. hole than an $1\frac{1}{8}$-in. one; therefore the round drift-bolt is at least 25 per cent more efficient per pound of metal than the square one.

The holding power of a 1-in. round bolt in a $1\frac{3}{8}$-in. hole in white pine, when drawn back immediately after driving, is a trifle over 10,000 lbs. per linear foot of bolt, a mean of 42 experiments on 7 pieces of timber. Twelve experiments on 3 sticks of Norway pine, under conditions similar to the preceding, gave 9000 lbs. per linear foot of bolt. Experiments upon 4 sticks of hemlock seem to show that the resistance is practically the same as white pine.

One-inch round screw-bolts were screwed into $1\frac{3}{16}$, $1\frac{1}{4}$, and $1\frac{3}{8}$-in. holes and immediately drawn back, the result being that there was but little difference for the different-sized holes. Half of the bolts had 8 threads to the inch and half had 12, the latter giving a very little the greater resistance. The resistance for the screw-bolts was about 50 per cent more than the maximum resistance of the plain round rods.

The report says: "Two classes of blunt points were used: Long, blunt points, tapered back for a distance of $1\frac{1}{2}$ to 2 in. and reduced to a round section, on square as well as round bolts, with a diameter less than that of the hole into which it was driven. They were pointed hot. Short, blunt points were reduced in size at an angle of about 45° by cold hammering, the point of the square bolt remaining square, with rounded corners, the intention being more to remove all cutting edges from the point than to reduce it much in size or change the square sections to round." The experiments were not so arranged as to make it possible to draw any reliable conclusion as to the relative merits of the two forms of points; but if the experiments show anything in this respect, it is that the resistance of bolts having "long, blunt points" is about ten per cent more than those having "short, blunt points."

Brooklyn Bridge Experiments.—Experiments made in connection with the construction of the East River Bridge by Mr. F. Collingwood and Colonel Paine, and communicated by the former, gave a holding power of 12,000 lbs. per linear foot of bolt for a 1-in. round rod driven into a $1\frac{3}{8}$-in. hole in first quality Georgia pine, and a resistance of 15,000 lbs. in a $1\frac{1}{4}$-in. hole. It was found that in lighter timber containing less pitch the holding power was about 20 per cent *less ;* and in very dense wood, containing *more* pitch, about 10 per cent more.

University of Illinois Experiments.—A third series of experiments was made by Mr. J. B. Tscharner in the testing laboratory of the University of Illinois, and published in full in " No. 4, Selected Papers of the Civil Engineers' Club of the University of Illinois." According to these experiments, the average holding power of a 1-in. round rod driven into a $1\frac{3}{8}$-in. hole in pine, perpendicular to the grain, is 6000 lbs. per linear foot ; and under the same conditions the holding power in oak is 15,600 lbs. per linear foot. The holding power of the bolt driven parallel to the grain is almost exactly half as much as when driven perpendicular to the grain. If the holding power of a 1-in. rod in a $1\frac{3}{8}$-in. hole be designated as 1, the holding power in a $1\frac{1}{4}$-in. hole is 1.69 ; in a $1\frac{3}{8}$-in. hole, 2.13 ; and in a $1\frac{3}{8}$-in. hole, 1.09. The holding power decreases very rapidly as the bolt is withdrawn.

Dowels.—In place of drift-bolts with point and head, plain iron bars, either square or round, are frequently resorted to. These are not forged or altered in any way, but are placed in the structure in just the condition that they are sheared from the rods, the only precaution taken being to see that they are straight.

The ties are frequently dowelled to the stringers. Pins made of $\frac{5}{8}$-in. round iron cut into pieces 5 in. long, are of a very good size. They weigh 0.4304 lb. each.

One method of fastening the posts, caps, and sills together is by means of dowels, $\frac{3}{4}$ in. by 8 in., which weigh about one pound each.

The following list gives the weight of one inch of a bar of iron of the various diameters most frequently employed in this kind of work :

1 inch square,	.	.	.	.	.	0.2806 lb.	$\frac{5}{8}$ inch diam. round,	.	.	.	0.1240 lb.
1 " diam. round,	.	.	.		.	0.2204 "	$\frac{5}{8}$ " square,	.	.	.	0.1096 "
$\frac{7}{8}$ " square,	.	.	.	.	.	0.2149 "	$\frac{1}{2}$ " diam. round,	.	.	.	0.0860 "
$\frac{7}{8}$ " diam. round,	.	.	.		.	0.1687 "	$\frac{1}{2}$ " square,	.	.	.	0.0701 "
$\frac{3}{4}$ " square,	.	.	.	.	.	0.1579 "	$\frac{1}{2}$ " diam. round,	.	.	.	0.0551 "

Bolts.—Bolts for holding the stringer-pieces together, fastening on the braces, guardrails, etc., are made of $\frac{3}{4}$-in. round iron. They vary in length of course, according to the use they are intended for. A head should be forged on one end, and a good, deep, well-formed right-hand thread cut upon the other for an appropriate distance. There are three kinds of heads in use in trestle-building: the round or button head, the flat countersunk head, and the ordinary square head (Fig. 129).

Square nuts with a thickness equal to the diameter of the bolt, and each side to twice the diameter, are the best. The outer top corners of the nuts and square heads should be chamfered. A cast-iron washer, from 3 in. to 3½ in. in diameter, is to be placed beneath both head and nut of all bolts. The bolts are driven through holes bored in the timber, and which should be $\frac{1}{16}$ in. less in diameter than the bolts, so as to insure a snug fit.

FIG. 129.
BOLTS.

While the weight of the bolt will be somewhat affected by the shape of the head, still the weight given in Table XIV may be used in making up preliminary estimates, as the error will be on the safe side ; i.e., too heavy.

TABLE XIV.

Approximate Weight of Bolts in Lbs., with Square Heads and Nuts, including both.

Length under Head in Inches.	Diameter in Inches.				
	½	⅝	¾	⅞	1
6	0.59	1.01			
7	0.64	1.10			
8	0.70	1.19			
9	0.75	1.27			
10	0.81	1.36	2.10	3.05	4.23
11	0.86	1.44	2.22	3.22	4.45
12	0.92	1.53	2.35	3.39	4.67
13	0.97	1.62	2.47	3.55	4.89
14	1.03	1.70	2.59	3.72	5.11
15	1.08	1.79	2.72	3.89	5.34
16		1.87	2.84	4.06	5.56
17		1.96	2.97	4.23	5.78
18		2.05	3.09	4.40	6.00
19			3.21	4.57	6.22
20			3.34	4.74	6.44
21			3.46	4.90	6.66
22			3.59	5.07	6.88
23			3.71	5.24	7.10
24			3.83	5.41	7.32

In ordering bolts the term " grip" is sometimes employed, meaning the total thickness of the material to be held together, or, in other words, the distance between the inside faces of the washers.

Lag-screws.—A lag-screw (Fig. 130) is little more than a very large wood-screw, with a square head similar to a bolt-head. A hole the full size of the shank should be bored through the first timber, otherwise the screw will not draw the timbers together. For the balance of the distance the hole should be bored much smaller. Under the head of each screw a wrought washer should be placed. The following table gives the details of the proper size of washer to use for different-sized lag-screws :

FIG. 130.—LAG-SCREW.

TABLE XV.

Proper Size of Wrought Washers.

Diam. Lag-screw.	Diam. of Washer.	Diam. of Hole.	Thickness Wire-gauge.	No. in 150 lbs.	Weight of one in lbs.
$\frac{1}{2}$ inch	$1\frac{3}{4}$ inches	$\frac{9}{16}$ inch	No. 12	4500	.0333
$\frac{5}{8}$ "	$1\frac{3}{4}$ "	$\frac{11}{16}$ "	" 10	3500	.06
$\frac{3}{4}$ "	2 "	$\frac{13}{16}$ "	" 10	1600	.0938

Separators, Thimbles, Packing Washers.—These were described when treating of stringers. They are made of cast-iron, which should be of good quality and free from blow-holes. Table XVI gives their dimensions and approximate weight.

TABLE XVI.

Details of Cast-iron Separators (see Figs. 62 to 72).

Kind.	Dimensions in Inches.						Weight in lbs.
	Diam. of Rims or Ends.	Thickness of Rim.	Breadth of Rim or Ends.	Thickness of Disk or Length of Spool from Outside Face to Outside Face of Ends.	Diam. of Hole.	Diam. of Spool or Smallest Diameter.	
Fig. 62	3		1	$\frac{1}{4}$	$\frac{7}{8}$		1.7
" 63	3		1	$\frac{3}{8}$	$\frac{3}{4}$		1.03
" 64	3	$\frac{3}{8}$	$1\frac{1}{4}$	$\frac{1}{4}$	$\frac{3}{4}$		1.5
" 65	$3\frac{3}{16}$		$1\frac{1}{4}$	$\frac{7}{8}$	$1\frac{1}{4}$	$2\frac{7}{16}$	0.6
" 66	$2\frac{1}{2}$	$\frac{3}{8}$	1	$\frac{9}{16}$	$\frac{3}{4}$		0.6
" 67	4		$\frac{5}{8}$	$4\frac{1}{2}$	$\frac{3}{4}$	2	5.5
" 68	4		$\frac{3}{4}$	3	$\frac{3}{4}$	2	3.25
" 69*	4		$\frac{5}{8}$	6	1	$1\frac{3}{4}$	1.7
" 70	4		$\frac{3}{8}$	6	$\frac{7}{8}$	$1\frac{3}{4}$	3.75
" 71	3		2	2	$\frac{3}{4}$	2	2.5
" 72	3		$\frac{1}{2}$	4	$\frac{3}{4}$	$1\frac{1}{4}$	1.75

* The six smaller holes are $\frac{5}{8}''$ in diameter.

Washers.—Cast-iron washers are used very extensively. They are always placed under the heads and nuts of all bolts in the structure. Fig. 131 gives a few of the designs in

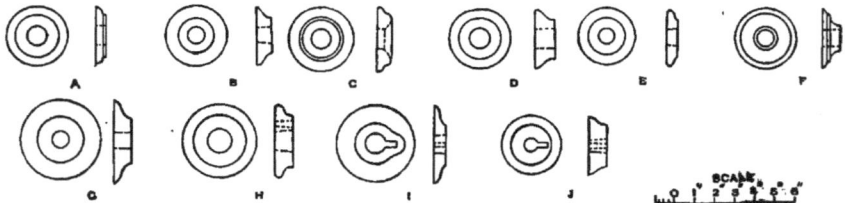

FIG. 131.—CAST-IRON WASHERS.

use, and Table XVII their weight and dimensions. The solid washers are placed under the heads of the bolts, and those having either a slot or second hole in them under the nuts. The purpose of these slots or holes is to enable a nail to be driven in close to the nut after it has been screwed down tight, to serve as a nut-lock.

TABLE XVII.

Details of Cast-iron Washers.

Kind. Fig. 131.	Dimensions in Inches.				Weight in lbs.
	Diam. of Back.	Diam. of Face.	Diam. of Hole.	Thickness.	
A	3	2 1/4	1	1/2	
B	3	1 3/4	7/8	3/4	
C	3 1/4	2 1/4	1	3/4	
D	3	2	1	1	
E	2 3/4	1 3/4	7/8	1/2	
F	3	1 1/8	7/8	3/4	
G	4 5/8	2 5/8	7/8	7/8	
H	3	2	3/4	3/4	
I	4	2	1	5/8	
J	3 7/8	2 1/8	1 3/16	3/4	
Similar to B	3 1/2	2	1	3/4	1.25
" " G	4 1/4	2	1	3/4	1.375

As wrought-iron washers are used to a greater or lesser extent in this class of work, a table giving the details of the standard washers as now manufactured, is appended.

TABLE XVIII.

Showing the Average Number of Wrought-iron Washers in a Keg of 150 lbs., of each Standard Size,

As adopted by "The Association of Bolt and Nut Manufacturers of the U. S."

Diameter.	Size of Hole.	Thickness Wire-gauge.	Size of Bolt.	No. in 150 lbs.
1/2	1/4	No. 18	3/16	80.000
5/8	5/16	" 16	1/4	34.285
3/4	5/16	" 16	1/4	22.000
7/8	3/8	" 16	5/16	18.500
1	7/16	" 14	3/8	10.550
1 1/4	1/2	" 14	7/16	7.500
1 3/8	9/16	" 12	1/2	4.500
1 1/2	5/8	" 12	9/16	3.850
1 3/4	11/16	" 10	5/8	2.500
2	13/16	" 10	3/4	1.600
2 1/4	15/16	" 9	7/8	1.300
2 1/2	1 1/16	" 9	1	950
2 3/4	1 1/4	" 9	1 1/8	700
3	1 3/8	" 9	1 1/4	550
3 1/2	1 1/2	" 9	1 3/8	450

These washers are merely circles stamped from sheet-iron, with a hole punched through the centre of them.

Nut-locks.—Special nut-locks are not required in trestle-work. The method of locking by driving a nail close to the side of the nut, through a hole in the washer, as mentioned when treating of cast washers, is as good and cheap a one as could be desired. Nicking the threads of the bolts with a centre-punch, after the nuts have been screwed home, is another very good way.*

* This method is used on the Texas & Pacific Railway.

CHAPTER VIII.

CONNECTION WITH EMBANKMENT—PROTECTION AGAINST ACCIDENTS.

Connection with Embankment.—There may be said to be two principal methods of connecting trestles with the embankment; viz., by sills built in the embankment itself, and by a pile-bent placed at its edge.

There are several ways of arranging the bank-sills. Sometimes they are piled up criss-cross, after the same fashion as in building a crib, several layers high. They should be of 12-in. × 12-in. timber, and at least 10 ft. long, and much better if the crosswise ones are 12 ft., securely fastened together by a drift-bolt wherever they cross each other. It is seldom that more than two sticks are used in each layer; those of the top layer should be at right angles to the centre line of the road, and placed quite close together over the centre of the crib. Their upper surfaces should be on the same grade level as the caps, so that the stringers will have a good bearing, the stringers being securely drift-bolted to them. After everything is in place earth should be packed in closely both inside and around the crib, and the bank carried out to at least the middle of the first bay. It will frequently be found necessary to protect the end of the bank from being washed away either by a revetment of logs, by sheet-piling, by rip-rap, or by other means.

Rather than arrange the bank-sills crib-fashion, some prefer to lay from two to eight or more pieces of the same size timber close together, on the same level and at right angles to the road. In this case, as before, the stringers should be drift-bolted to the bank-sills.

With whichever arrangement is used, however, the bank should be allowed to stand as long as possible before putting in the bank-sills, so that it will have time to settle.

The preferable way to connect the trestle with the bank is by a bank-bent. This is either a pile-bent of three or four piles, or a light framed bent. In any case the ends of the stringers are usually protected from contact with the earth by a piece of heavy plank nailed across them, called a dump-board. Plate VI, Part II, shows a form of bank-bent. It sometimes happens that it is necessary to plank up behind the bank-bent so as to prevent the embankment spreading beneath the trestle. In this case, if a pile-bent is used, it should be strongly built, and the piles penetrate to a considerable depth, especially if the bent be of any height. It is also well to brace the tops of the piles against the foot of the piles in the next bent, so as to prevent the bank-bent being forced over by the pressure of the embankment behind it. If a framed bent is chosen, it should be strong and heavy, and well braced against its neighbor, both diagonally and by girts acting as struts. If possible, the girts or horizontal bracing should extend clear across the whole structure, be of heavy material, have butted joints and be well fastened, so as to avoid buckling; in other words, they should fulfil all of the requirements for struts.

Rerailing Guards.—All extensive and all high trestles should be protected by a rerailing guard, and it would be far better if *all* the trestles were, without regard to their size.

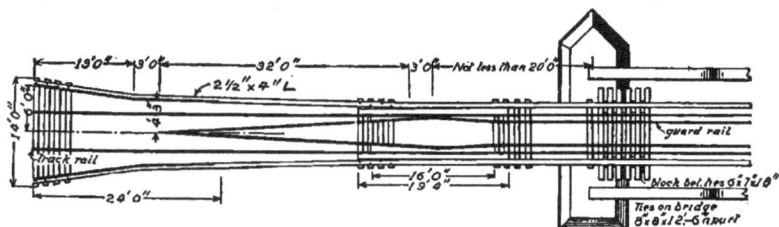

Guard Rail and Rerailer.

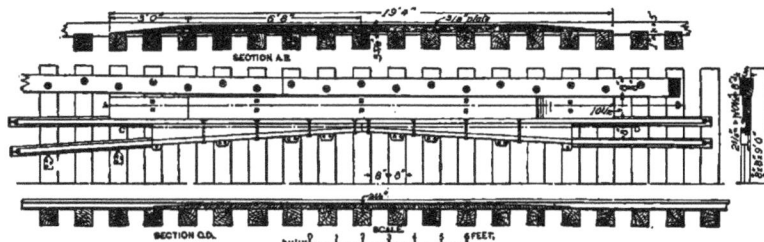

Plan and Sections of Rerailer.

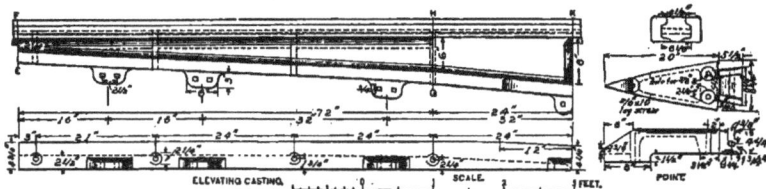

Elevating Casting and Point.

Section on G H.   Section on E F.

Section on J K.

FIG. 132.—LATIMER BRIDGE-GUARD.

If this cannot be done, then collision-posts at least should be erected to guard them. Even where rerailing guards are used it is an excellent plan to supplement them by collision-posts arranged so as to stop a car, the truck of which has moved half of the gauge or more out of line. This would at least save the bridge, even though it would not prevent an accident. Fig. 118 shows such collision-posts.

In Fig. 132, the details of the Latimer bridge-guard, as used on the Savannah, Florida & Western and the Charleston & Savannah Railways, Mr. B. W. Howe, Jr., Chief Engineer, are given.

Refuge-bays.—On all trestles of any length, say two hundred feet or over, refuge-bays or small railed platforms to receive workmen or track-walkers who may be caught on the bridge by a train should be placed every two or three hundred feet apart. These cost but very little, and are very efficient in insuring greater safety to employees, especially on single-track trestles.

Fig. 133 shows an excellent attachment for this purpose.

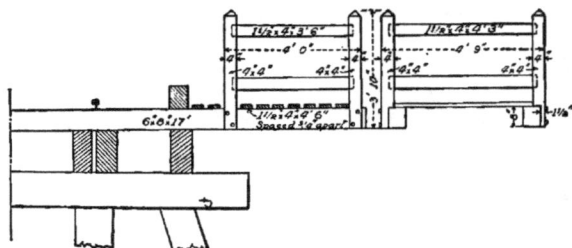

Fig. 133.—Refuge-bay.

Every fourth or fifth refuge-bay on trestles over one thousand feet long, especially when on or approached by a curve, should be made large enough to receive the hand-car; and when the section-men or the repair-gang are at work on the bridge they should always be compelled to place the hand-car on the refuge-bay, together with all idle tools, before they begin work.

Foot-walks.—Some engineers recommend the laying of foot-walks, composed of three or four rows of 4-inch plank, along the centre of the trestle. This, however, for a number of reasons, does not seem desirable, even though it make the life of the track-walker more endurable. Among the objections may be mentioned :

1st. A tendency to make the track-walkers and others careless in their examination of the structure.

2d. It offers a greater temptation to people to make a highway of the trestle on account of the greater comfort and ease with which it may be crossed, and hence encourages the public to trespass upon the railroad company's property, and that upon the most dangerous places.

3d. It increases, very largely, the area for cinders from the engines to fall upon, and hence makes the risk of fire much greater.

Fire Protection.—As long as wooden trestling is used fire will be one of the most troublesome subjects to deal with. There are several devices, which are now employed more or less extensively, to reduce the danger from this source.

The one most extensively used is to place tubs or half-barrels, which are kept full of water, at short intervals along the trestle. They should never be over two hundred feet apart, and should each be supplied with a pail or generous-sized dipper. The pails should never be made of wood, as they are liable to be found in anything but a serviceable condition when most needed. Both "Indurated fibre" and "Granite" or enamelled iron-ware are excellent materials for this purpose. The water in the tubs should never be allowed to become low, and it should be the imperative duty of the track-walker to see that they are kept full. Common kerosene oil-barrels cut in half make very good tubs. On single-track trestles these are placed on one side upon the ends of two ties, which are purposely made longer than the others for this use. On double-track trestles they are placed between the two tracks. As this safeguard is very cheap indeed, there is no reason why every trestle in the country, without exception, should not be so protected. In the colder portions of the country there is, of course, the disadvantage of the water freezing in winter, but this is no reason for depriving the public of what little benefit there is in the apparatus during the balance of the year. Railroad companies, for their own sake, should adopt it, as it would frequently lessen the cost of an accident by furnishing immediate means for the extinguishment of many a fire in its incipiency, after a wreck has occurred.

A second method is to cover the stringers with a strip of sheet-iron about three or four inches wider than they are, before placing the ties, etc., on them. See Plate III, Part II. Common sheet-iron of about No. 27 gauge is very good for this use. The iron should be protected from rust by some means. A good preventive is common tar. Before putting the iron in place it should be warmed, and thoroughly painted all over with the hot tar.

A third kind of fire protection is that illustrated in Plate VIII, Part II, in which the trestle has a solid floor which is covered with earth.

Not only should means be provided to prevent the spread of and to put out fires that have once started from unavoidable or accidental causes, but every precaution possible to prevent them approaching from the outside should also be taken. The right of way to a width of 15 to 20 ft. from either side of the trestle should be kept perfectly clear of all combustible matter of any kind at all times. Not only should this rule be closely observed, but no amount of any moment should be allowed to accumulate outside of this limit. Within it, all weeds and tall grass should be kept closely cut. When construction or repairs are going on, all chips and small blocks should be raked up in a heap at the close of the day, at a safe distance from the work, and set on fire. If the work is being done by contract, the contractor ought to see that this is done for his own protection. When the trestle is on a line in course of construction, the right of way should be thoroughly cleared, the necessary space grubbed, and the rubbish cleaned up and completely burned before erection is allowed to be begun, or at any rate before the trestle is accepted by the engineer or the contractor estimated for the work done. Any trees off of the right of way which are likely to fall upon and injure the trestle should be felled. It is the railroad company's place to obtain permission to do this, though the contractor may be rightly called upon to do the work for which he may be estimated at the same rate as for clearing.

CHAPTER IX.

FIELD ENGINEERING AND ERECTING.

THERE are several methods of laying out the ground preparatory to erecting a trestle. Of course the exact method of procedure will depend, to a certain extent, upon the surrounding circumstances.

The centre-line should be run in carefully with a transit, and the stakes, which should be well made and stout, driven firmly into ground. A stake should be placed on the centre-line at each bent, and a tack, located by the instrument, driven in.

For a pile-trestle on land the instrument is set up over each centre stake and the proper angle turned off, and stakes driven in on either side at the proper places for the outside piles. The tape is then stretched between the centre and outside stakes, and stakes marking the position of the inside piles driven in. Some prefer, for framed bents especially, to use hubs in place of stakes, and centre a tack on each one. This, however, is an unnecessary refinement. For framed bents it is preferable to place the stakes, which should be driven down pretty close to the surface in this case, a foot ahead of the centre of the bent. A centre stake and one a little distance out on either side is all that is necessary. A mark is made on the sill half-way between the two vertical posts, and when the bent is put in position this mark is placed opposite the centre stake. Care is taken to see that the sill sets back the proper distance from all of the stakes,—6 in. between the stake and the face of the sill in the case mentioned. Of course when a framed bent has a pile-foundation the piles are located in the same manner as for pile-trestles. When the foundation is of masonry the centre-line in both directions is first laid out and then stakes driven in in such a manner that when strings are stretched between them they mark the outline of the top of the masonry. A mark or stake giving the elevation of the top is also given. After the foundation is in, the centre is marked on top of it.

For use on this kind of work a 50-foot tape is much more convenient than a chain. An ordinary linen tape, so thoroughly coated with paint that it will not stretch much, is accurate enough, though some prefer a metallic tape. A steel tape is by no means necessary, as some younger engineers are inclined to think, and is very liable to be broken.

It is exceedingly convenient to have a bench-mark, the elevation of which is somewhere near grade, within one or two hundred feet of either end of the trestle, so that it may be easily seen through the level from the end of the embankment. The elevation of the top of the bent can be given with the instrument while the bent is being put in place, or a bench can be established at the end of the bank, and the foreman can then obtain the elevation with an ordinary carpenter's level and straight-edge, allowance being made by him for the grade. In the latter case the work should always be checked, every day or so, with the Wye level.

After the bents are completed and in place the centre-line is to be marked on each cap by a nail or tack, so that the stringers may be placed in their proper positions. Track centres are

56

given, of course, in a similar manner as on the grading, after the ties have been placed in position and the structure otherwise completed.

When the trestle is over water and on a tangent there are several ways of lining in the piles. An instrument may be used, but as a rule this is not necessary unless the trestle is very long. A less expensive way is to place very long stakes, standing four or five feet out of the ground, on line with the rows of piles, having two sets of stakes, one fifty or one hundred feet behind the other, and have the foreman line the piles in with these. One edge of the stakes should be on the line, of course, instead of having the line pass through them. The outside stakes should also be driven at the proper batter. As the work progresses these stakes may be replaced by narrow boards nailed to the piles. The results should be checked by the engineer in charge, from time to time.

Erecting.—The method used in erection depends upon the location. Where it is permissible the bents are generally framed together while lying upon the ground, with the sill so placed that when the bent is raised it will be in its proper position. They are raised, usually by blocks and a fall, the rope being drawn in by a horse-power or steam hoisting-engine or by a gang of men. As soon as the bents have reached the upright position they are fastened to those already erected by temporary bracing, which should be supplemented by the permanent longitudinal bracing as rapidly as possible, if such is to be used. If not, then the stringers should be placed in position. Stay-ropes should be attached to the bent before it is raised, so that when it reaches its upright position it cannot be pulled over. Of course when the bents are of any considerable height they are liable to considerable racking if erected in this manner. Attaching an additional fall to a couple of timbers lashed to the bent about half-way up, one timber on either side, tends to prevent this to a considerable extent. However, great care must be taken to draw in the ropes of the two falls at the proper rates. It is in this part of the work of building high trestles that the third class of high-trestle structures proves so convenient. The lowest deck is erected and the purlins placed upon it. Then the timbers for the bents of the next deck are put together on a temporary staging formed by placing a flooring on the purlins. These bents are then erected the same as though they were upon the ground, the purlins put on top of them, and the same process carried on as before until the full height is reached. Then there is less liability to injury or loss, while in the course of erection, through the bents falling from lack of temporary bracing, as is too frequently the case.

Another method is to complete the work and lay the track as rapidly as the bents are placed. The bents, in this case, may either all be framed at any convenient place, or on the ground as before, and then brought to and placed in position or raised by a derrick and hoisting machine placed on a flat car. The boom of the derrick should, as a rule, be long enough to reach out so as to place the second bent beyond the completed work in position. The bents for a trestle much over 15 ft. high could not, of course, be conveniently carried any distance.

Sometimes the bents are built in place. This method is absolutely necessary for very high structures, unless they are of the type of class three. The cost is generally greater than with the previous methods. One of the strong arguments advanced for both cluster-bent and compound timber trestles is the economy with which they may be erected by this method,

On account of the smaller size of the timbers they may be handled with much greater ease and rapidity.

TOOLS.

The following is a description of the tools used by the carpenters in trestle-building. Most of them will be found absolutely indispensable; the remainder greatly facilitate the work:

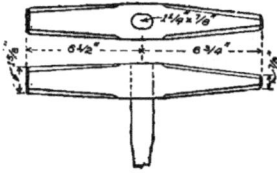

FIG. 134.—SPIKE-MAUL.

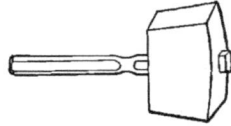

FIG. 135.—MALLET.

Hammers.—Practically the only hammers used to any extent are the spike-mauls. Fig. 134 gives the details of a good maul.

Mallets.—Mallets are merely wooden hammers. They are used principally to drive the chisels into the wood. Being of wood, they do not, of course, injure the handles of the chisels as steel hammers would. They are made either of a wood called lignum vitæ or of hickory. The former is more durable, and also more costly. Fig. 135 shows one form of mallet very commonly used.

FIG. 136.—CROSS-CUT SAW.

Saws.—A cross-cut saw, such as is shown in Fig. 136, about five feet long, is exceedingly useful. If enough work is laid out beforehand, so that two men can be assigned to the saw and be kept constantly employed, great economy will result. If the men have to stop between cuts to lay out the work themselves, more or less time is lost in making the change and hunting for the tools, and it often happens that one man remains entirely idle while the other is preparing the work.

FIG. 137.—HAND-SAW.

Hand cross-cut saws are also required (Fig. 137). These should be of the heavier patterns, and the blade at least two feet long. If the handle is bound with brass and at right angles to the back, so that the saw may be used as a square, it will be found to be very convenient. These saws are used for the lighter parts of the work, such as notching the ties, guard-rails, ends of stringers, etc.

In addition to these it will be necessary to have some rip-saws. These are used for sawing with the grain of the wood, and are about the same size as hand cross-cut saws, or a little larger. The teeth are larger and differently shaped than those of the cross-cut saw.

Boring-machines.—For boring out mortises preparatory to finishing with the chisel, a boring-machine is exceedingly economical and useful.

Ship Augers.—For boring holes for bolts, drift-bolts, lag-screws, etc., a ship auger, such as is shown in Fig. 138, is most commonly used. Augers of this style should be long enough to enable a man to use them standing without having to stoop.

FIG. 141.—BROADAXE.

FIGS. 139, 140.—AXES.

FIG. 138.—SHIP AUGER.

FIG. 142.—HATCHET.

Axes.—A common long-handle axe (Figs. 139 and 140) is very useful. They are made of different weights; usually, each man has his own particular liking in this regard. About 4 lbs. is a good weight for the head, exclusive of the helve. A 5-lb. axe is rather heavy, while one weighing only 3 lbs. is rather light.

In addition to the common axe, broadaxes (Fig. 141) and hatchets (Fig. 142) are found convenient.

Adzes.—An adze may be defined as an axe with the cutting edge set at right angles to the handle. This tool, which is absolutely necessary to economical and rapid work, is shown in Fig. 143.

FIG. 143.—ADZE.

FIG. 144.—FRAMING CHISEL.

Chisels.—The best form of chisels for this kind of work is the firmer or framing chisel (Fig. 144). The handle should be held in a socket forged on the upper end of the blade, and should have its top end protected by an iron ring. The most convenient widths are $1\frac{1}{2}$ in. and 2 in.

They are used to cut out mortises, and the notches in the ties, guard-rails, etc. Tanged chisels are of no use, as the work is too heavy for them.

Squares and Rules.—The ordinary steel framing square, made of sheet steel about $\frac{1}{8}$ in. thick, and with one arm about two feet and the other about twelve inches long, in addition to a batter template, such as is shown in Fig. 45, is all that is required in this line. The arms should be graduated in inches and quarter-inches.

Besides the common two-foot rule, it greatly facilitates matters to have a strip of board about $\frac{1}{2}$ in. thick by $2\frac{1}{2}$ in. wide and 10 ft. long, divided into feet and numbered both ways, one set of numbers being in red and the other in black, and separated from each other by a line through the centre of the stick. The first foot of either set should be divided into inches and quarters of an inch.

Cant-hooks and Lug-hooks.—Both cant and lug hooks will be found necessary and useful in handling the timber. Fig. 145 shows a cant-hook and Fig. 146 a modification of a

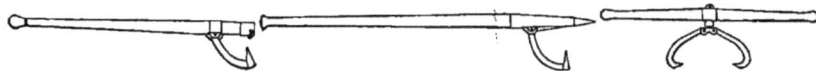

FIG. 145.—CANT-HOOK.   FIG. 146.—PEAVEY.   FIG. 147.—LUG-HOOK.

cant-hook called a peavey. A lug-hook is shown in Fig. 147.

Log-wheels.—A pair of log-wheels will be found very useful for carrying the timber from one place to another. They are merely two strongly-built wheels, of a large diameter, with a broad, heavy tire, united by a very strong axle. To the axle is attached a shaft, so that a team of horses or yoke of oxen may be hitched to the wheels. The wheels are backed over one end of a timber to be moved, and the end raised from the ground by means of chains and an arrangement of levers. The rear end of the stick is allowed to drag upon the ground.

Wrenches.—For trestle-building, the ordinary monkey-wrench is of little use. As the nuts are all of one or two sizes, the form shown in Fig. 148 is one of the most convenient.

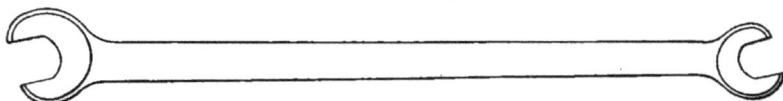

FIG. 148.—WRENCH.

Another form in common use is made upon the same principle as a clock key. This enables the men to tighten up many of the nuts without stooping.

Hoisting-machines.—Under the head of ERECTION, hoisting-machines were spoken of as being used to aid in raising the bents and timbers. Whether horse or steam power machines are used will depend on several conditions, among which may be mentioned the extent of the work and the means for the transportation of the machine to the site. A horse-power machine can be much more easily transported, and can be carried over roads over which it would be either impossible to transport a steam machine or prohibitory in cost. Work cannot be prosecuted as rapidly, of course, with a horse-power as with a steam machine.

Saw-mills.—It sometimes happens that a very extensive piece of trestling will be needed in a location where there is plenty of timber, but no saw-mills at hand. In this case, if it is deemed necessary to use sawn timber, a portable saw-mill will be found very convenient. These mills are generally arranged so that they can be very conveniently and easily moved from place to place, and may be obtained of various capacities.

CHAPTER X.

PRESERVATION OF JOINTS AND STANDARD SPECIFICATIONS.

Preservation of Joints.—Wherever two surfaces of timber touch, they should always be painted with some preservative material. White-lead is sometimes used, but is rather costly. Common tar heated very hot, coal-tar, and creosote oil are excellent for this purpose, while they also have the advantage of cheapness. Of course, if all of the timber could be treated with a preservative agent so much the better; but this is generally too expensive. Creosoted timber is probably as durable as that preserved by any other process. This subject of timber preservation, however, will not be treated of in this work. Those who may wish to know more about the subject are referred to the various books and papers on the subject.*

Standard Specifications.—The degree of care used, and the completeness of the specifications drawn up by different roads, varies between exceedingly wide limits. Some say almost nothing on the subject at all, only, perhaps, devoting one or two lines in the General Specifications to the subject; others draw up a special set entirely devoted to the subject.

The following set of specifications were compiled from the best parts of the best standards in use that could be obtained. The paragraphs having the same headings are alternative paragraphs, any one of which may be used to suit the special conditions of the road.

STANDARD SPECIFICATIONS FOR WOODEN TRESTLES.†

CLEARING.

Before commmencing work on any structure, the ground must be entirely cleared of logs, brush, and trees for the entire width of the right of way. All material of a combustible nature must be placed in piles at convenient places, and completely burned.

---

* Short and interesting articles on this subject appeared in the *Railroad Gazette* of Sept. 5, 1890, and Sept. 19, 1890.

† The form used on the Cincinnati, New Orleans & Texas Pacific Ry. and associate roads (Cincinnati Southern Ry., Alabama Great Southern R. R., New Orleans & North-Eastern R. R., Vicksburg & Meridian R. R., Vicksburg, Shreveport & Pacific R. R.), G. B. Nicholson, Chief Engineer, has been very closely followed. The specifications of the following roads have also been drawn from:

Central Railroad & Banking Co. of Georgia, V. H. Kriegshaber, Asst. Engineer.
Ohio Connecting Ry., M. J. Becker, Chief Engineer.
Georgia Pacific Ry. Co., I. Y. Sage, General Superintendent.
Cleveland, Akron & Columbus Ry. Co.
Gulf, Colorado & Santa Fé Ry. Co., B. F. Booker, Asst. Engineer.
St. Paul, Minneapolis & Manitoba Ry. Co., N. D. Miller, Chief Engineer.
Florence R. R., F. Gardner, Chief Engineer.
Brantford, Waterloo & Lake Erie R. R.
Specifications for Standard Pile and Timber Trestle Bridging,—*Eng. News.*
French Broad Valley R. R., H. M. Ramseur, Chief Engineer.

Dangerous trees, liable to fall on the trestle, when outside the right of way, must be felled by the contractor; it being understood that the railroad company is to obtain permission from the land-owner.

Such portion of the right of way, as may be deemed necessary by the engineer, shall be grubbed.

DRAWINGS.

The drawings are to the scale indicated and marked; but in all cases the figures are to be taken, and in case of omission the engineer in charge is to be referred to for dimensions. Under no circumstances are the drawings to be scaled either by the contractor or by any of his men. The Engineer will be required to mark the dimensions upon the contractor's blue print, and to keep a record of the same in his office.

DIMENSIONS.

All posts, braces, clamps, stringers, packing-blocks, ties, guard-timbers, sills, and all timber generally, will be of the exact dimensions given and figured upon the plan. Variations from these will only be allowed upon the written consent of the engineer in charge.

TIMBER.

All timber shall be of good quality and of such kinds as the engineer may direct, free from wind-shakes, wanes, black, loose, or unsound knots, worm-holes, and all descriptions of decay. It must be sawed true and out of wind, and full size. Under no circumstances will any timber cut from dead logs be allowed to be placed in any portion of the structure: it must in every case be cut from living trees.

PILES.

Piles shall be of good live ———

They will be either round of square, as may be required by the engineer.

Round piles must be straight, and have all the bark peeled off. They must have at least twelve (12) inches diameter of heart at the cut-off, when cut to grade to receive the cap. The smaller end must be at least eight inches in diameter.

Square piles must be hewn (or sawed) twelve (12) inches square. Each pile must have at least nine (9) inches of heart wood on each face, from the head of the pile, after being cut to grade, to five feet below the surface of the ground in which the pile is driven.

All piles must be properly pointed. They shall, if required, be shod with cast or wrought iron shoes, made according to the plan furnished by the engineer. In driving they shall be capped with suitable wrought-iron rings, if necessary, to prevent splitting. The actual cost, delivered on the ground, of the necessary shoes and rings will be allowed the contractor.

They must be driven to hard bottom, or until they do not sink more than five inches under the last five (5) blows by a hammer of at least 2000 lbs. weight falling twenty-five (25) feet. A heavier hammer with a shorter fall is preferred.

All piles injured in driving, or driven out of place, shall either be cut off or withdrawn, as the engineer may elect, and another one driven in its stead. The pile thus replaced will not be paid for.

All piles under track-stringers must be accurately spaced and driven vertically, and in each bent the batter-piles will be driven at the angle shown.

Piles shall be measured by the lineal foot after they are driven and cut off, and the price per lineal foot shall be understood to cover the expenses of transportation, driving, cutting off, removing the bark, and all labor and materials required in the performance of the work, but that portion of each pile cut off shall be estimated and paid for by the lineal foot as " Piles cut off."

The contractor must give all facilities in his power to aid the pile-recorder in his duties.

Parts of pile-heads projecting beyond the cap must be adzed off to a slope of 45 degrees.

FRAMING.

All framing must be done to a close fit, and in a thorough and workmanlike manner. No blocking or shimings of any kind will be allowed in making joints, nor will open joints be accepted.

All joints, ends of posts, piles, etc., and all surfaces of wood on wood shall be thoroughly

painted with
$\left\{\begin{array}{l} \text{hot creosote-oil and covered with a coat of hot asphaltum,} \\ \text{hot asphaltum,} \\ \text{hot common tar,} \\ \text{a good thick coat of pure white-lead ground in and mixed with} \\ \qquad \text{pure linseed-oil.} \end{array}\right\}$

All bolt and other holes bored in any part of the work must be thoroughly saturated with

* $\left\{\begin{array}{l} \text{hot creosote-oil,} \\ \text{hot asphaltum,} \\ \text{hot tar,} \\ \text{coal-tar,} \\ \text{white-lead mixed with} \\ \qquad \text{linseed-oil.} \end{array}\right\}$ And all bolts and drift-bolts before being put in place * $\left\{\begin{array}{l} \text{warmed and coated with hot creosote-oil,} \\ \text{warmed and coated with hot asphaltum,} \\ \text{warmed and coated with hot tar,} \\ \text{coated with coal-tar,} \\ \text{coated with white-lead and linseed-oil.} \end{array}\right.$

must be

All bolt-holes for bolts three quarters ($\frac{3}{4}$) of an inch in diameter or over must be bored with an auger one eighth ($\frac{1}{8}$) of an inch smaller in diameter than the bolt, in order to secure a perfectly tight fit of the bolt in the hole. For bolts five eighths ($\frac{5}{8}$) of an inch in diameter or smaller the auger must be one sixteenth ($\frac{1}{16}$) of an inch smaller for the same reason.

TRESTLES ON CURVES.

Where any trestle-bridge is built on a curve the blocking for the elevation of the outer rail, or other means for elevating the outer rail, will be as per standard drawings for the same, a copy of which will be furnished from the Chief Engineer's office.

---

* Alternative methods of treatment.

CREOSOTED TRESTLES.

All piles used in creosoted trestles must have the bark peeled off, and be pointed, before treatment. None of the sap wood must be hewn from the piles. No notching or cutting of the piles will be allowed after treatment, except the sawing off of the head of the pile to the proper level for the reception of the cap, and the levelling of such part of the head as may project from under the cap.

The heads of all creosoted piles, after the necessary cutting and trimming has been done to receive the cap, must be saturated with hot-creosote oil, and then covered with hot asphaltum before putting the caps in place.

Timber in creosoted trestles must be cut and framed to the proper dimensions before treatment. No cutting or trimming of any kind will be allowed after treatment, except the boring of the necessary bolt-holes.

Hot creosote-oil must be poured into the bolt-holes before the insertion of the bolts, in such a manner that the entire surface of the holes shall receive a coating of creosote-oil.

TREATMENT OF CREOSOTED PILES AND TIMBER.

All creosoted timber and piles shall be prepared in accordance with the following process:

The timber and piles, after having been cut and trimmed to the proper length, size, and shape, shall be submitted to a contact steaming inside the injection-cylinders, which shall last from two to three hours, according to the size of the timbers; then to a heat not to exceed 230°, F., in a vacuum of twenty-four (24) inches of mercury, for a period long enough to thoroughly dry the wood. The creosote-oil, heated to a temperature of about 175°, shall then be let in the injection-cylinder and forced into the wood under a pressure of 150 pounds per square inch, until not less than fifteen (15) pounds of oil to the cubic foot of wood has been absorbed.

The oil must contain at least ten (10) per cent of carbolic and cresylic acids, and have at least twelve (12) per cent of naphthaline.

IRON.

Wrought-iron.—All wrought-iron must be of the best quality of American refined iron, tough, ductile, uniform in quality, and must have a limit of elasticity of not less than twenty six thousand (26,000) pounds per square inch.

All bolts must be perfect in every respect, and have nuts and screws to the full standard sizes due to their diameters. The thickness of the nut shall not be less than the diameter of the bolts and the size of its square not less than twice the diameter of the bolt.

The heads of all bolts shall be $\left\{ \begin{array}{l} \text{square} \\ \text{countersunk} \\ \text{round button} \end{array} \right\}$ heads,

with a thickness not less than the diameter of the bolt, and the size of its square not less than twice the diameter of the bolt.

with a thickness at the centre of not less than three quarters of the diameter of the bolt, and an extreme diameter of not less than two and one half times the diameter of the bolt.

countersunk on the under side so as to fit into a cup-washer, with an extreme diameter of not less than twice the diameter of the bolt.

Cast-iron.—All castings must be from good, tough metal, of a quality capable of bearing a weight of five hundred and fifty (550) pounds, suspended at the centre of a bar one (1) inch square, four and one half (4½) feet between supports. They must be smooth, well-shaped. free from air-holes, cracks, cinders, and other imperfections.

All iron, before leaving the shop must be thoroughly soaked in boiled linseed-oil.

INSPECTION AND ACCEPTANCE.

All materials will be subject to the inspection and acceptance of the Engineer before being used. The Contractor must give all proper facilities for making such inspection thorough.

Any omission to disapprove of the work, by the Engineer, at the time of a monthly or other estimate being made, shall not be construed as an acceptance of any defective work.

PROTECTION AGAINST FIRE.

The Contractor must each evening, before quitting work, remove all shavings, borings, and scraps of wood from the deck of the trestle, and from proximity to the bents or piles, and on the completion of the work must take down all staging used in the erection, and burn all shavings, chips, etc., and remove all pieces of timber to a distance sufficient to insure safety from fire.

ROADS AND HIGHWAYS.

Commodious passing places for public and private roads shall be kept in good condition by the Contractor, and he shall open and maintain thereafter a good and safe road for passage on horseback along the whole length of his work.

RUNNING OF TRAINS.*

The Contractor shall so conduct all his operations as not to impede the running of trains or the operation of the road. He will be responsible to the Railroad Company for all injuries to rolling-stock or damage from wrecks caused by his negligence. The cost of such damage will be retained from his monthly and final estimates.

RISKS.

The Contractor shall assume all risks from floods, storms, and casualties of every description, except those caused by the Railroad Company, until the final acceptance of the work.

LABOR AND MATERIAL.

The Contractor must furnish all material and labor incidental to or in any way connected with the manufacture, transportation, erection, and maintenance of the structure until its final acceptance.

Disorderly, quarrelsome, or incompetent men in the employ of the Contractor, or those who persist in doing bad work in disregard of these specifications, must be discharged by the Contractor when requested to do so by the Engineer.

---

* This clause may be inserted when the repairs or renewals are made by contract, or on the double-tracking of a road.

Whenever the Chief Engineer may deem it advisable, he may name the rates and prices to be paid by the Contractors, for such time as he may designate, to the several classes of laborers and mechanics in their employ, and for the hire of horses, mules, teams, etc., and these shall not be exceeded ; and having given due notice to the Contractors of his action in regard to these matters, they shall be bound to obey his orders in relation thereto. The Chief Engineer shall not, however, name a rate or price for any class of labor, etc., higher than the maximum rates being paid by the Contractor paying the highest for that class.

INTOXICATING LIQUORS.

Contractors will not themselves, nor by their agents, give nor sell any intoxicating liquors to their workmen, or any persons at or near the line of the railway, nor allow any to be brought on the works by the laborers or any other person, and will do all in their power to discountenance their use in the vicinity of the work by persons in their employ. A continued disregard for this clause will, if deemed necessary by the Engineer, be considered as a good and sufficient reason for declaring the contract forfeited.

DAMAGES AND TRESPASS.

Contractors shall be liable for all damages to landholders, arising from loss or injury to crops or cattle, sustained by any cause or thing connected with the works, or through any of their agents or workmen. They will not allow any person in their employ to commit trespass on the premises of persons in the vicinity of the works, and will forthwith, at the request of the Engineer, discharge from their employ any that may be guilty of committing damage in this respect. They will also maintain any fences that may be necessary for the proper protection of any property or crops.

REMOVAL OF DEFECTIVE WORK.

The Contractors will remove at their own expense any material disapproved by the Engineer ; and will remove and rebuild, without extra charge, and within such time as may be fixed by the Engineer, any work appearing to the Engineer, during the progress of the work or after its completion, to be unsoundly or improperly executed, notwithstanding that any certificate may have been issued as due to the execution of the same. The Engineer shall, however, give notice of defective work to the Contractors as soon as he shall become cognizant of the same. On default of the Contractors to replace the work as directed by the Engineer, such work may be done by the Railroad Company at the Contractors' expense.

DELAYS.

No charge shall be made by the Contractor for hindrances and delay, from any cause, in the progress of any portion of his work ; but it may entitle him to an extension of the time allowed for completing the work sufficient to compensate for the detention, to be determined by the Engineer, provided he shall give the Engineer in charge immediate notice, in writing, of the detention.

EXTRA WORK.

No claim shall be allowed for extra work unless done in pursuance of a written order from the Engineer, and the claim made at the first estimate after the work was executed, unless the Chief Engineer, at his discretion, should direct the claim, or such part of it as he may deem just and equitable, to be allowed.

Unless a price is specified in the contract for the class of work performed, extra work will be paid for at the actual cost of the material remaining in the structure after its completion and the cost of the labor for executing the work, plus fifteen (15) per cent of this total. This fifteen (15) per cent will be understood to include the use of and cost of all tools and temporary structures, staging, etc., and the Contractor's profit, and no extra allowance over and above this will be made.

INFORMATION AND FORCE ACCOUNTS.

The Contractor will aid the Engineer in every way possible in obtaining information, and freely furnish any which he may possess, by access to his books and accounts, in regard to the cost of work, labor, time, material, force account, and such other items as the Engineer may require for the proper execution of the work, and shall make such reports to him from time to time as he may deem necessary and expedient.

PROSECUTION OF THE WORK.

The Contractor shall commence his work at such points as the Engineer may direct, and shall conform to his directions as to the order of time in which different parts of the work shall be done, as well as the force required to complete the work at the time specified in the contract. In case the Contractor shall refuse or neglect to obey the orders of the Engineer in the above respects, then the Engineer shall have power to either declare the contract null and void and relet the work, or to hire such force and buy such tools at the Contractor's expense as may be necessary for the proper conduct of the work, as may in his judgment be to the best interests of the Railroad Company.

CHANGES.

At any time during the execution or before the commencement of the work the Engineer shall be at liberty to make such changes as he may deem necessary, whether the quantities are increased or diminished by such changes, and the Contractor shall be entitled to no claim on account of such changes beyond the actual amount of the work done according to these specifications at the prices stipulated in the contract, unless such work is made more expensive to him, when such rates as may be deemed just and equitable by the Chief Engineer will be allowed him ; if, on the other hand, the work is made less expensive, a corresponding deduction may be made.

QUANTITIES.

It is distinctly understood that the quantities of work estimated are approximate, and the Railroad Company reserves the right of having built only such kinds and quantities, and according to such plans, as the nature or economy of the work may, in the opinion of the Engineer, require.

ENGINEER.

The term Engineer will be understood to mean the Chief Engineer, or any of his authorized Assistants or Inspectors, and all directions given by them, under his authority, shall be fully and implicitly followed, carried out, and obeyed by the Contractor and his agents and employees.

PRICE AND PAYMENT.

The prices bid will include the furnishing of materials, tools, scaffolding, watching, and all other items of expense in any way connected with the execution and maintenance of the work, until it is finally accepted and received as completed.

The Contractor will only be paid for the piles, timber, and iron left in the structure after completion.   No wastage in any kind of material will be paid for except in the case of piles, when the "piles' cut-off," and which cannot be used on any other part of the Contractor's work, will be paid for at the rate agreed upon.   After the material cut off is paid for it is to be considered as the property of the Railroad Company, and is neither to be removed nor used by the Contractor without the consent of the Engineer, and then only upon the repayment of the price which has been paid for it.

The piles and "piles' cut-off " will be paid for by the lineal foot, the former to be driven and in place.

The timber and lumber will be paid for by the thousand feet, board-measure, remaining in and necessary to the completed structure.

The iron will be paid for by the pound actually remaining in the structure after its completion.

The masonry for foundations will be paid for by the cubic yard.

The excavations for foundations will be paid for by the cubic yard.

The retained percentage will not be paid on the cost of any single structure until the final estimate is due on the entire work embraced in the contract.

If the building of the trestle is let with the contracts for grading or under a general contract, then many of these clauses may be omitted, as they are merely general requirements applicable to all classes of work.   Many of the clauses would also be omitted or changed somewhat under the different conditions existing in different sections of the country.   The effort has been made, however, to make them as generally applicable and as concise as possible, and all of the clauses inserted have been selected on account of their general excellence and justice to both Contractor and Railroad Company.

A form of proposal is as follows :

THE RED RIVER RAILROAD COMPANY.

PROPOSAL FOR BUILDING TRESTLES.

The undersigned hereby certify that they have *personally and carefully examined the location and the plans and specifications for the trestles* on the first, second, and third divisions *on the line of the Red River Railroad.*

Having made such examinations, the undersigned hereby propose to the said Red River Railroad Company to furnish all the material and do all the work required for the construction and completion of said first, second, and third division *trestles, in accordance with said specifications and plans, and upon the acceptance of this proposal do hereby bind* themselves *to enter into and execute a contract for the same at the following*

PRICES:

Material.	Unit.	Approximate quantities. May be more or less.	Rate.	
Foundation excavation—Earth, ...	Cubic yard.	25		22
Solid rock, . . .	Cubic yard.	15		90
Foundation masonry, ...	Cubic yard.	56	5	00
Round white-pine piles, not creosoted, . .	Lineal foot.	1500		35
" " " " " " cut off,	Lineal foot.			00
. " oak " " "	Lineal foot.			
" " " " " cut off,	Lineal foot.			
Etc. etc. etc.				
Square yellow-pine piles, not creosoted, . .	Lineal foot.	1000		35
" " " " " cut off,	Lineal foot.			08
Etc. etc. etc.				
White-pine timber, not creosoted, erected, .	M. B. M.	100 M.	30	00
Oak " " " "	M. B. M.	10 M.	40	00
Etc. etc. etc.				
Round oak piles, creosoted, ...	Lineal foot.	1725		75
" " " " cut off, ...	Lineal foot.			30
Etc. etc. etc.				
White-pine timber, creosoted, erected, . .	M. B. M.	750 M.	30	00
Etc. etc. etc.				
Wrought-iron, ...	Pound.	10,000		04
Cast-iron, ...	Pound.	1250		02½
. . . .				
. . . .				
. . . .				

The undersigned further propose to commence work within ten *days from date of contract, and to complete the same within* sixty *days from date of contract.*

Signed this sixth *day of* January, *1890.*

Name of Firm, SMITH BROS. & COMPANY.

By { GEO. H. SMITH.
       WM. R. SMITH.
       ED. C. BROWN.

Post Office address of Contractor :

97 Great George Street,

New York City,

New York.

This form in blank, for filling out, should be printed and bound with the specifications, together with the agreement or contract. Those portions of the form printed in Roman type are left blank for filling in by the bidders, excepting in the table of prices, where only the prices are left blank.

CHAPTER XI

BILLS OF MATERIAL, RECORDS, AND MAINTENANCE.

ONE of the most perplexing duties to the young engineer is, perhaps, the making out of proper bills of materials for trestle-work. The following is an example of a properly made out bill of material:

TRESTLE NO. 6.

DIVISION No. 2; RESIDENCY No. 4.

BILL OF TIMBER.

No. of Bent.	Number of Pieces.	Name.	Size.	Feet B. M.	Total feet B. M.
1—Height, 9 feet.					
	2	Cap.	6″ × 12″ × 14′ 0″	168	
	2	Plumb-posts.	12″ × 12″ × 8′ 0″	192	
	2	Batter-posts.	12″ × 12″ × 9′ 0	216	
	1	Sill.	12″ × 12″ × 12′ 4″	148	
	8	Blocks—Mud-sills.	12″ × 12″ × 2′ 6″	240	964
2—Height, 13 feet.					
	2	Cap.	6″ × 12″ × 14′ 0″	168	
	2	Plumb-posts.	12″ × 12″ × 12′ 0″	288	
	2	Batter-posts.	12″ × 12″ × 13′ 2″	316	
	1	Sill.	12″ × 12″ × 14′ 2″	170	
	2	Sway-braces.	3″ × 10″ × 16′ 6″	83	
	8	Blocks—Mud-sills.	12″ × 12″ × 2′ 6″	240	1265
3—Height, 10 feet.					
	2	Cap.	6″ × 12″ × 14′ 0″	168	
	2	Plumb-posts.	12″ × 12″ × 9′ 0″	216	
	2	Batter-posts.	12″ × 12″ × 10′ 1″	242	
	1	Sill.	12″ × 12″ × 12′ 8″	152	
	2	Sway-braces.	3″ × 10″ × 14′ 0″	70	
	8	Blocks—Mud-sills.	12″ × 12″ × 2′ 6″	240	1088
Floor System and Miscellaneous Parts:					
	8	Bank-sills.	12″ × 12″ × 12′ 0″	1152	
	10	Stringers and Jack-stringers.	8″ × 16″ × 25′ 0″	2667	
	4	Stringers.	8″ × 16″ × 12′ 6″	534	
	51	Ties.	6″ × 8″ × 12′ 0″	2448	
	9	Guard-rails.	6″ × 8″ × 20′ 0″	720	7521
Grand Total, ...					10,838

BILL OF IRON.

No. of Pieces.	Name.	Use.	Size.	Weight.
		Wrought Iron.		
24	Drift-bolts.	Stringers to Bank-sills.	¾″ sq. × 24″	
26	Drift-bolts.	Stringers to Caps.	¾″ sq. × 24″	
6	Drift-bolts.	Sills to Mud-sills.	¾″ sq. × 20″	
102	Boat-spikes.	Ties to Stringers.	½″ × 12″	
150	Boat-spikes.	Guard-rails to Ties.	½″ × 12″	
26	Bolts.	Guard-rails to Jack-stringers.	¾″ × 31½″	
12	Bolts.	Caps to Posts.	¾″ × 22″	
16	Bolts.	Sway-bracing.	¾″ × 20″	
32	Bolts.	Packing for Stringers.	¾″ × 22″	
		Total,		
		Cast Iron.		
172	Washers.	Under heads and nuts of Bolts.	1″ × 3″	
32	Separators.	Between Stringers.	2″ × 3″	
		Total,		

Signed, WILLIAM BOSS, *Resident Engineer.*

Jan. 25, 1890.

A copy of all such bills as these should be made in a letter-book. In making out the estimates of timber in feet, B. M., the contractor should always be allowed the full size of any stick between the extreme ends of the tenons, and where the ends or tenons are required to be cut on a skew, the full size for the length with square ends required to cut the piece.

The following is the rule for finding the number of feet, B. M., in any stick of timber, or in lumber one inch or over in thickness :

Multiply the breadth and thickness in inches together, and divide by twelve. Multiply this result by the length in feet and fractions of a foot, and the final result will be the number of feet, B. M., in the stick.

Putting this in the form of an algebraic expression, we have

$$\text{Feet B. M.} = \frac{b \times t \times L}{12}.$$

$b =$ breadth in inches ;

$t =$ thickness in inches (when one inch or over) ;

$L =$ length in feet and fraction of a foot.

When the lumber is less than one inch in thickness it is always counted as though it were a full inch thick.

It will be found that if such a table as that shown below be made out for bents up to a moderate height, varying by six inches, and blue prints of it sent to the different resident and division engineers, considerable labor and time will be saved, and many annoying, and at times serious, errors avoided.

NORTH AMERICAN RAILROAD COMPANY.

BILL OF TIMBER FOR STANDARD TRESTLES.

PILE.							PILE.						
Height from surface of Ground to top of Cap.		Caps 6″ x 12″		Sway-braces 3″ x 10″		Intermediate Caps 3″ x 10″	Feet B. M.	Height from surface of Ground to top of Cap.		Caps 6″ x 12		Sway-braces 3″ x 10″		Intermediate Caps 3″ x 10″	Feet B. M.
Ft.	Ins.	Pcs.	Length Ft. Ins.	Pcs.	Length Ft. Ins.	Pcs.	Length Ft. Ins.		Ft.	Ins.	Pcs.	Length Ft. Ins.	Pcs.	Length Ft. Ins.	Pcs.	Length Ft. Ins.	
5	0	2	14 0				168	8	0						
5	6	2	14 0				168	8	6						
6	0	2	14 0					9	0						
6	6	2	14 0					9	6						
7	0							10	0	2	14 0	2	16 6		251
7	6							10	6						
								11	0						

FRAMED.

Height from bottom of Sill to top of Cap.		Caps 6" x 12"			Plumb-posts 12" x 12"			Batter-posts 12" x 12"			Sill 12" x 12"			Sway-braces 3" x 10"			Intermediate Caps 3" x 10"			Mud-sills 12" x 12"			Feet B.M.
Ft.	Ins.	Pcs.	Length. Ft.	Ins.	Pcs.	Length. Ft.	Ins.	Pcs.	Length. Ft.	Ins.	Pcs.	Length. Ft.	Ins.	Pcs.	Length. Ft.	Ins.	Pcs.	Length. Ft.	Ins.	Pcs.	Length. Ft.	Ins.	
5	0	2	14	0	2	4	6	2	4	6	1	16	0							8	2	6	816
5	6																						
6	0																						
6	6																						
7	0																						
7	6																						
8	0																						
8	6																						
9	0																						
9	6																						
10	0																						
10	6																						
11	0																						

General.—Each trestle will require :

```
    8 Bank-sills 12 in. X 12 in. X 12 ft................ 1152 ft. B. M.
    4 Dump-boards 4 in. X 8 in. X 11 ft. 4 in.........  121  "   "   "
    Stringer-pieces 8 in. X 16 in. X 25 ft., contain each.  267  "   "   "
    Ties 6 in. X 8 in. X 12 ft.          "      "   ...   48  "   "   "
    Guard-rails 6 in. X 8 in. X 20 ft.    "      "   ...   80  "   "   "
```

Many other devices for furnishing aid in making out bills of material have been invented, and are used to a greater or lesser extent in the various offices throughout the country. One of the most notable of those which have come to the Author's notice, and of which the originator is not known, is that of drawing a bent to a large scale,—say three-quarters of an inch to the foot, on paper that will not vary much with changes in the atmosphere. The sills for heights varying by regular amounts—six inches is very good— are then drawn in. When the length of any part, for any height of bent, is needed, it can readily be scaled directly from the drawing.

Construction records in detail are, of course, made out each month for all of the trestle-work built since the previous estimate was taken up. The thoroughness and completeness of these records vary considerably on different roads.

Following are some forms of construction records as kept on the Norfolk & Western Railroad. These records are very complete, and are to be recommended.

Trestle No........
Section No........
Residency No........

NORFOLK & WESTERN RAILROAD COMPANY.—CONSTRUCTION DEPARTMENT.

MASONRY EXHIBIT OF TRESTLE-FOOTINGS.

..............18..

..............Contractor.

..............Resident Engineer.

STATION.	No. of Bent.	Sub-grade.	Sub-grade to top of Masonry.	GROUND.			ELEVATIONS.							LENGTH OF FOOTING.			Distance between Footings.	Cubic Yards.	Show cross-section of irregular and stepped footings, also skew-bents.
							TOP OF MASONRY.				BOTTOM OF MASONRY.									
				Left.	Centre.	Right.	Left.	Centre.	Right.	Distance from centre.	Left.	Centre.	Right.	Distance from centre.	Left.	Centre.	Right.			

This sheet is 13½ in. long by 8½ in. wide, and is ruled horizontally in blue ink, with five lines to the inch.

It is indorsed on the back for filing as follows:

NORFOLK & WESTERN RAIL-ROAD.

CONSTRUCTION DEPARTMENT.

MASONRY EXHIBIT OF TRESTLE-FOOTINGS.

Trestle No........

Section No........

Estimate No........

.................18..

.................*Contractor.*

.................*Cubic Yards.*

This indorsement is so placed that it will be on the outside when the sheet is folded into four parts across its length so as to make a bundle about 3⅜ in. by 8½ in.

NORFOLK & WESTERN RAILROAD CO.

Timber Estimate No. . . . .**18.**
Trestle No. . . . . *Section No.* . . . . *Res.* . . . .*Contractors.*
East End of Stringers, Sta. . . . . *West End, Sta* . . . .*Length* . . . .*feet.*

STATION.	No. of Bent.	SHOULDER TO SHOULDER.							OVER ALL.						
		Plumb Post.		Outside Batter Post.		Inside Batter Post.			Sill.		Diagonal Bracing.		Longitudinal Bracing.		
		No.	Length.	No.	Length.	No.	Length.		No.	Length.	No.	Length.	No.	Length.	
Add Tenons,															
Totals,															

SUMMARY.	Dimensions.	Feet B. M.		Engineer.
	Sizes.		
	Length.		
	Description of Pieces.	Plumb Posts, Outside Batter Posts, " Inside " " Sills, Caps, Intermediate Caps, " " Corbels, " Knee Braces, Straining Beams, " Caps, Diagonal Bracing, " Longitudinal Bracing. " Waling Strips, " Wall Plates, Stringers, Packing Pieces, " " Ties, Guard Rails, Mud Blocks,	
	No.		

This sheet is of the same size and indorsed upon the back in the same manner as the foregoing one. The indorsement is as follows:

NORFOLK & WESTERN RAILROAD COMPANY.

TRESTLE ITEM SHEET.

Trestle No......
Section No......
Estimate No......
......18
Contractor.
Ft. B. M......
Total Amount, $......
Engineer.

Now, if no more than one trestle is put upon any one sheet, the sheets may be bound in their proper order upon the completion of the road, and will then form excellent records for the use of the Maintenance of Way Department. These blanks are, of course, filled out and signed by the Resident Engineer, and their summaries entered upon the "Detail Item Sheet" of the Residency for the corresponding month. These should then be forwarded to the Division Engineer, who compiles the following report from those of the several Residencies under him:

NORFOLK & WESTERN RAILROAD COMPANY.

Estimate No........ ....*Division.* ....

..................18 BRIDGE, TRESTLE AND TIMBER ESTIMATE. *Contractor.*

Res. Sec.	DESIGNATION OF STRUCTURE.	CHARACTER OF STRUCTURE.	IRON BRIDGES. Length Out to Out.	WOODEN BRIDGES. Length Out to Out.	TIMBER DELIVERED. Pine.	TIMBER DELIVERED. Oak.	TIMBER FRAMED.	TRESTLE ERECTED.	IRON, BOLTS, ETC.	REMARKS.

TOTAL TO DATE
PREVIOUS ESTIMATE NO.
CURRENT ESTIMATE.

This sheet is 7 in. by 17 in., and is intended to be folded once each way. **One half of the** back is ruled for a summary as follows :

SUMMARY.

Items.	Previous Estimate.	Current Estimate.	Total to Date.	Rate.	Amount.

Certified by . . . .

. . . . *Engineer.*   Total Amount.

On one half of the remaining half of the back is the following indorsement for filing purposes :

NORFOLK & WESTERN R. R. CO.

BRIDGES, TRESTLE AND TIMBER ESTIMATE.

Estimate No. . . . . 18 . . . . .

. . . . *Contractor.*

	Dollars.	Cts.
Total Amount, - - -		
Retained Percentage, -		
Balance, - - - - -		
Former Payments, - -		
Amount Due, - - -		

After the road has been finished and turned over to the operating and maintenance departments, inspections of the trestles, the same as with all the other properties, should be frequently and regularly made. As to the frequency of these special inspections the practice and opinions vary. A personal inspection of all the structures by the Engineer of Maintenance of Way or of Bridges and Buildings should be made at least once a year at an auspicious season. On the New York, Lake Erie & Western Railroad this inspection is required twice a year. Of course inspections of single structures should be made at any time when the necessities of the case demand them. It is good practice, where there is any considerable amount of trestling and bridging on a division, to have a competent inspector whose sole business is to inspect and oversee repairs to the structures. He should personally and carefully examine every structure under his charge once every month, or two months, as the location of the road may require, and report their condition on proper blanks to the Division Engineer or Division Superintendent. These officers, in their turn, after examining and approving these reports, should forward them to the Engineer of Maintenance of Way, or the Engineer of Bridges and Buildings, as the case may be. Every part of each structure should be carefully and critically examined from all sides, and the inspector should be required under all circumstances to examine trestles, not only from their deck, but also from beneath. Proper facilities should be afforded him for this purpose. In urgent cases he should report by telegraph or letter from the nearest station, as the matter may require. In addition to this the track-walkers should keep a constant watch upon all trestles, and report their condition daily to the inspector. A pad ruled as follows will be found very convenient for the track-walkers to make these reports upon:

NORTH AMERICAN RAILROAD COMPANY.

TRACK-WALKER'S DAILY REPORT ON THE CONDITION OF BRIDGES AND TRESTLES.

NUMBER OF BRIDGE.	TIME.		CONDITION.	
	A.M.	P.M.	A.M.	P.M.
150	10:30	1:40	X	X
151	9:45	3:15	X	X
152				
153				
154				
155				

.... *Track-walker.*

....18....

Six inches wide by seven inches long, with fifty sheets in a pad, is a very good size. It is also advisable to have a cardboard cover which will shut over the face of the pad in the same manner as the cover of a book.

The sheets are folded across for filing, with the following indorsement on one half:

NORTH AMERICAN RAILROAD COMPANY.

TRACK-WALKER'S DAILY REPORT.

BRIDGES AND TRESTLES NOS. 150 TO 155.

................, ...18....

.... *Track-walker.*

On the other half of the back these instructions are printed:

You will carefully examine each bridge and trestle over which you may pass, and enter their *condition* in the proper column and on the proper line of this blank.

You will also enter the *time* of such examination in the proper column and on the same line.

X in the column headed "Condition" means "all right."

O means injured, or unsafe by *fire, washout*, or other means.

These reports *must* be forwarded every evening to the Inspector of Bridges.

A report must be made out every time a bridge or trestle is passed over, even if three or four times a day.

In case of O, such fact must be *telegraphed at once* from nearest office to the *Inspector of Bridges*, the *Division Engineer*, and the *Division Superintendent*.

A repeated *disregard* of these instructions will be considered a sufficient cause for *discharge*.

An axe or hatchet and a small auger are absolutely indispensable to an inspector for the proper performance of his duty. Frequently the soundness of a piece of timber can be tested by pounding upon it with a hammer, and listening to the sound which the blows make. In case of any question, a hole should be bored into the timber with the auger. This will, of course, settle the matter beyond all doubt. These holes should always be filled up immediately after boring them, either by driving a plug in very tightly or with putty It is not advisable to bore many holes in one piece of timber, as they greatly weaken it. A better way to arrive at a just conclusion, it seems, is to drive in a long, thin nail, such as a wire-nail. The degree of ease with which the nail penetrates the wood is a very good test of its condition. In pile-trestles on land, where the foot of the piles can be reached, it is very good practice to dig away the ground around them, for a foot or so in depth, once every twelve or eighteen months, in order that the least durable part may be inspected. After the inspection the earth is replaced and properly tamped. This inspection need not be begun until after the pile has been in the ground for several years. The length of time which it is advisable to allow to elapse before beginning it will depend largely upon the kind of timber. Thorough records of all these inspections should be carefully made and preserved. In order to be able to properly and definitely locate any part of any structure beyond the question of a doubt, all of the bridges, trestles, etc., should be numbered consecutively, beginning at one end of road and going toward the other. Then in each trestle the bents should be numbered in the same direction ; the stringers, guard-rails and longitudinal bracing should also be numbered from right to left ; the ties over each bay should be numbered ; and finally, the stories beginning at the top and going to the bottom, should be treated similarly. By this means any one acquainted with the system of the road can take a description of any part from the Bridge Book. and locate it upon the ground beyond all question.

A heading suitable for the inspector's reports and records is as follows :

NORTH AMERICAN RAILROAD COMPANY.

MAINTENANCE OF WAY DEPARTMENT.

TRESTLE INSPECTION.

Bridges No. — to No. —

Bridge Number.	Date, 18—	Number of Bent.	Class.	Piles.				Foundation.		Story.	Sill and Intermediate Cap.	Posts.				Cap.	Stringers.						Ties.														Guard-rails.		Sway-braces.		Longitudinal Bracing.				Iron.	Condition of Fire Protection.	Remarks.
				1	2	3	4	Class.	Condition.			1	2	3	4		1	2	3	4	5	6	1	2	3	4	5	6	7	8	9	10	11	12	13	14	1	2	1	2	1	2	3	4			

Symbols.—O.K. = good for more than one year.

D = dangerous, and must be removed at once.

6 = must be removed within six months.

S = must be removed within one year.

X = replaced since last report.

I hereby certify that I personally made the above examinations and reports on the dates named, and that they are true and correct in every particular, to the best of my knowledge and belief.

(Signed)

————, Inspector.

————— 18—

Examined and approved,

————, Division Engineer.

————, Division Superintendent.

————— 18—

Every space opposite a bent should be filled in to show that that part has been examined. If there is no such part, a line should be drawn through the space. The filling in of the report should be done on the ground. After the reports are sent to the Engineer of Maintenance of Way they may be placed in proper order and bound. These reports should be forwarded at regular and stated intervals, and a copy of them kept by the Division Engineer.

They should be on paper about 24 in. long by 12 in. wide, padded in lots of one hundred sheets, in a similar manner to the Daily Reports of Track-walkers. Two or three blotting-sheets should be attached to the front cover, and the filling-in required to be done in ink, a fountain or stylographic pen being useful for this purpose. The instructions and affidavit may either be printed at the bottom or on the back of the sheet, as is found most desirable.

The following is the order of inspection required upon the Plant system of railways in South Carolina, Georgia, and Florida : *

Number of bents, piles, sills, legs, caps, corbels, chords, posts, braces, stringers, floor-beams, condition of cross-ties. Do piles in this bridge or trestle settle ? If so, state condition of shims, number of feet of standard guard-timber, condition.

Is the opening subject to wash at end, or at bottom ?

Total length of bridge.   Longitudinal braces every   bent, size, condition.

Are abutments protected by rock, revetment timbers, or any other protection ?   Condition of such protection.

A certificate as to the truth of the inspection is then given by the inspector.

The inspector should have direct charge of all bridge-repair gangs working on his division, and all orders or instructions to the men transmitted through him, and he be held responsible for their proper execution.

The following are the instructions issued by the New York, Lake Erie & Western Railroad, and the Burlington, Cedar Rapids & Northern Railroad, in regard to bridge-work :

BRIDGE INSPECTION ON THE ERIE.†

Under the system of inspection and reports now in force on the Erie there are employed 10 inspectors who report directly to their respective Division Roadmasters each month, and every three months the Division Roadmasters make inspections and report to the Division Superintendents. These quarterly reports are forwarded through the General Roadmaster and General Superintendent to the Engineer of Bridges and Buildings, in whose office they are filed. The inspectors have no other duties than those of inspection.

The blank forms for the inspectors' reports are on sheets 20 inches square. The heading of the sheet is as follows :

Form X. 402.   N. Y., L. E. & W. R. R. CO.

REPORT SHOWING CONDITION OF BRIDGES ON THE........DIVISION, MONTH ENDING........18....

The sheet is ruled in 17 vertical columns, of which the headings are as follows : Number ; Kind of Bridge, Wooden or Iron ; *General Conditions of* Masonry, Bed Plates, Rollers and Frames, Pedestals, Main Trusses or Girders, Lateral System. Iron Floor System, Rivets, Hangers, Castings, Paint ; Action under Trains ; Date of Inspection ; Remarks and Recommendations. This is to be signed by the Inspector, and at the foot is further space for remarks. On the back of the sheet are printed the following orders and directions :

NEW YORK, LAKE ERIE & WESTERN R. R. CO.

GENERAL ORDERS FOR THE INSPECTION OF BRIDGES.

1. Besides the constant and careful examination of all bridges by the regular Inspector, each Roadmaster shall make a personal and thorough inspection of the same once every three months.

2. A regular report of the condition of every bridge shall be made by the Inspector to the Division Roadmaster every month, upon blanks furnished for that purpose.

* Maj. C. S. Gadsden, *Railroad Gazette*, 1888, p. 652.   † *Railroad Gazette*, July 29, 1887.

3. The quarterly examination made by the Roadmaster shall be reported upon form No. X, 402 A, but must be signed by him and forwarded to the Division Superintendent, who shall in turn transmit it through the proper channels to the Engineer of Bridges and Buildings.

4. The Engineer of Bridges and Buildings will also make a stated personal examination of all bridges twice a year, besides the customary inspection of special cases, as reported from time to time, and upon the request of the General Superintendent.

5. The condition of the different parts of the bridges must be briefly stated under the appropriate heads on the blanks furnished, and in case of need, further information shall be given in the column of "Remarks and Recommendations."

6. Special reports by letter or telegraph, according to the urgency of the case. must be made by the Inspector or Roadmaster wherever any fault or defect is discovered that may, in their judgment, endanger the safety of the bridge.

7. All ordinary repairs, such as tightening loose rivets and renewing wooden floors on iron bridges, or replacing such parts of wooden structures as have become defective by age and are necessary for the safety of the bridge, shall be done without special orders.

8. When, however, alterations, additions, or expensive renewals of any bridge are contemplated and become necessary, they must be reported to the Engineer of Bridges and Buildings, who will then prepare the necessary plans and estimates for approval.   R. H. SOULE, *General Manager.*

DIRECTIONS, GIVING THE MOST IMPORTANT POINTS TO BE OBSERVED WHILE INSPECTING BRIDGES.

Masonry.

1. Each pier and abutment should be carefully looked over, especially those that have already given signs of yielding, either by settling in their foundations or by bulging from the pressure of the embankment they sustain.

2. Examine closely all pedestal stones, looking for cracks or evidence of crushing; they must be maintained level and firmly bedded upon the bridge-seats.

3. Keep the latter clean and free from all rubbish and cinder or coal, especially around the iron bed-plates.

Iron Bridges.

5. Examine carefully all pedestals, bed-plates, and rollers and their frames. The bed-plates should be perfectly level, the rollers should move freely and their axes should always be kept at a right angle to the line of the bridge. The pedestals should be free from all cracks and flaws, and have a uniform bearing upon all the rollers or upon the bed-plate at the fixed end.

8. In the main trusses look most closely at all the tension members, the rods and bottom chords, especially where they are composed of more than one member. If perfect, they should all be equally strained per square inch in any one panel, and when they are not, when one member is slack and the other tight, the case should be reported at once. The compression members, that is, the posts and top chords, should be straight, without a bend or bulge, and all the joints should bear closely against each other. The counter rods ought never to be allowed to hang loose, but they must not be adjusted while a load is upon the bridge, and they must not be tightened more than just enough to get a good bearing.

6. *All hangers, by which floor-beams or stringers are suspended, must constantly receive the closest attention.* Their bearing around the pins should always be equal and uniform over half the circumference of the latter. If the hangers are made of round or square iron they must be examined with great care in the semicircle where they are bent around the pins, and where flaws or fracture are most likely to occur, and it is of the utmost importance that the nuts on the ends of such hangers supporting the whole floor of the bridge are never permitted to become loose. A white streak painted across the face of the nut and its bearing will make it easy to detect at once any motion in the nut.

7. *The places where stringers are riveted or otherwise fastened to the floor-beams,* and which are generally not easy of access for inspection, on account of the wooden floor over them, *must be frequently and thoroughly examined.* Here the rivets are most likely to get loose, and the webs and flanges of the beams and stringers are more liable to fail from shearing or crushing than anywhere else.

8. The lateral systems and sway-bracing must never be neglected when a bridge is inspected. All the rods should be tight but not overstrained, as the struts are liable to be crippled if too much power is used in adjusting the tension members.

9. Cast-iron parts of all bridges, more particularly when in top chords or in joint boxes, must be closely examined. Should any cracks or breaks be discovered the fact must be at once reported. A hole of $\frac{1}{4}$ in. diameter if drilled at the end of a crack will frequently stop its extending further.

10. Riveted work should frequently be sounded with a hammer to detect loose rivets; and if they cannot be tightened at once their number and location must be reported on the monthly report.

11. No water must be allowed to collect in the interior of any cast or wrought iron parts; drain-holes should be kept open for that purpose, and must be provided if they do not exist.

Wooden Bridges.

12. After a wooden bridge or trestle has been in use over three years, a close inspection must be made twice a year as to the condition of the timber, by boring holes in suspicious-looking places, especially near the bridge-seats and at the ends of stringers and braces. The nature of the boring will reveal the fact if the timber is sound or decaying. Whenever splices exist in bottom chords, and principally in long-span bridges where they generally occur in every panel, it is very important to examine them thoroughly and to note if they are pulling apart, which would indicate a weakness or a defective clamp. The braces and counterbraces should always have a square and even bearing upon the angle-blocks, and the sliding away from their true position, if any, would be sure evidence that the bridge needs immediate adjustment.

13. Tubs filled with water and buckets should be kept constantly on hand on every span of all wooden bridges.

General Conditions.

14. The action of a bridge under a passing train is the best practical test of its stability, and no inspection shall be completed without having made such observation, and without having carefully noted any undue deflection, swaying or twisting of the bridge as a whole or any part thereof.

15. The Roadmaster should carefully measure with an instrument the absolute deflection and swaying of any bridge reported to him by the Inspector as doubtful, and if the movements are excessive must report the fact at once.

16. The tracks on the bridges as well as on the approaches thereto should always be in good line and surface; they should be firmly bedded on the approaches, so as to avoid any undue shock when a train comes on a bridge at a high rate of speed.

C. W. BUCHHOLZ,
Engineer of Bridges and Buildings.

This form is folded, and on the outer fold is the indorsement for filing under the proper division and date.

The form for the report of the quarterly inspection of the Roadmaster is precisely the same, except that the sheet is 20 in. × 25 in., to give room for three columns of remarks. These columns bear the headings, By Roadmaster. By Div. Superintendent, and by Gen'l Roadmaster, and are signed by these officers respectively, and the whole is signed by the General Superintendent when examined, approved, and forwarded by him to the Engineer of Bridges and Buildings.

INSTRUCTIONS TO BRIDGEMEN ON THE BURLINGTON, CEDAR RAPIDS AND NORTHERN R. R.; H. F. WHITE, CHIEF ENGINEER.†

INSTRUCTIONS TO BRIDGEMEN.

1. You will be furnished with the bill of material needed for each structure before your men are sent out on the work. You must, as soon as you reach the bridge site, check the material delivered with your bill to see that both agree, and must personally ascertain, as quickly as possible, if the bill of material includes all that will be required. If the requisite amount of material is not delivered, you must notify the Master Builder of the deficiency, that the same may be forwarded promptly. Any material unfit or not proper for the structures will be reported without delay, that another kind may be substituted.

2. You must see before starting for work that you are fully equipped with the necessary tools to do your work. You must bear in mind that you are liable to be called away at any time from work upon which you may be employed, to that of a more pressing nature, and in order that you may be fully prepared for such exigencies must see that you have the facilities at hand for moving from place to place, at short notice, and are provided, as far as practicable, with the necessary tools to do all kinds of bridge work.

3. All bridgemen are expected to be prompt at the depot when it is necessary for them to take trains to reach their work; as far as possible, they will be expected to board near the place where they work. Repeated failures to be in time for trains will be considered good grounds for dismissal; men so left will be docked for time in transit when they take the next train, and receive pay only for time actually at work.

4. All men in the service of the company must report to the head of their department any misconduct or negligence affecting the interests or safety of the road or property, and which may come within their knowledge. The withholding of such information to the detriment of the company's interests will be considered a proof of negligence and indifference to the company's interests.

---

† *Railroad Gazette*, Sept. 21, 1883.

5. Foremen must actively engage in their work with their men, and see that all the force working under their orders faithfully perform their duties and work full time.

6. Bridgemen will be held responsible for all company tools and material put in their charge. In case of breakage or loss, the company reserves the right to withhold from money now or hereafter due them, a sufficient amount to repair or replace them, as may be thought best by the head of department.

7. You must fill out in full all blanks and forward the same in accordance with instruction given, and must inform yourself about all rules and regulations of the company, and be governed accordingly in the prosecution of your work, and must study and always have a copy of the time-card in force.

Hand-cars must not be left on the track when not in use, but must always be safely cared for.

Signals must always be put out at the proper distance when the roadway is not in good condition for the passage of trains.

8. You must see that your men are not unnecessarily exposed to accidents which will in any way render the company liable for damages.

9. Bridgemen are expected to pay their own board promptly, and in case of failure, the company reserves the right to withhold from money now or hereafter due them a sufficient amount to pay the same, but does not assume any responsibility for board. A repetition of the offence will be considered sufficient cause for dismissal.

10. Tools must not be carried into the ladies' car, and employés of the bridge department must not occupy seats when by so doing passengers are obliged to stand.

11. Any employés not disposed to comply with these instructions are requested to leave the employ of the company at once. The orders will be read to or by each man employed before he commences work. Any failure to have this done will subject the Foreman to discharge from service.

12. Bridgemen, in cases of necessity, will be expected to work on Sunday, at the same rate as paid for work done on the other days of the week.

It is the practice on some roads to give premiums to the bridge-foreman who puts in his timber at the least cost per thousand feet, board measure. On the Charleston & Savannah Railroad * the practice is as follows:

"General Order 188, paragraph 9, provides for a premium for bridge-foremen.

"At the end of each three months the bridge-foreman who shall have put his timber in at the least cost per thousand feet, B. M., will be rewarded with a premium of fifteen ($15) dollars. At the same time a premium of ten ($10) dollars will be given to the bridge-foreman who shall have made the next best showing.

"The conditions of these premiums are as follows:

"(A) Only the actual time devoted to bridge-work will be considered, and fifteen (15) minutes will be allowed for each train passing during working hours.

"(B) All timber put in will be considered.

"(C) The work done must be strictly workmanlike, and in accordance with the standard plans."

As to the wisdom of adopting this premium system on all roads, it would be difficult to determine. Whether, in many instances, it might not lead to the slighting of work, where it would be difficult or impossible to discover it, is a very serious question.

The tools required in repair-work are much the same as those for building purposes. In addition to those described, a claw-bar, for drawing spikes, drift-bolts, etc., will be necessary. A small hydraulic jack will frequently be found very serviceable.

For the purpose of designating the bridges and other similar structures, as spoken of in the first part of this chapter, bridge-numbers, as they are called, are used. These are generally made of pieces of 2-in. plank attached to the bridge near one end by $\frac{5}{8}$-in. by 4-in. lag-screws with wrought-iron washers. Two forms of bridge-numbers are illustrated in Figs. 149 and 150.

---

* Maj. C. S. Gadsden, Supt. Chas. & Sav. R.R., on "Care of Trestles;" *Railroad Gazette*, 1888, p. 652.

The boards should be planed and painted white with several coats of white-lead, or, better still, zinc-white, ground in good linseed-oil. The figures (Fig. 150) are black, three

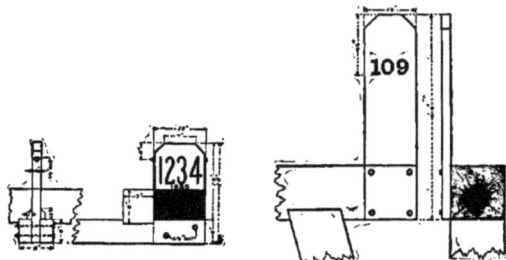

FIGS. 149, 150.—BRIDGE-NUMBERS.

inches in height, with the base fourteen inches below the top of the board. The numbers should be placed on the bridges with uniformity, i.e., they should occupy the same relative position on all the structures. For example, the following is the rule for placing them on the Atlantic & Pacific Railroad :

Position.—East side of right-hand end of cap on bent, fifteen feet from last or initial end of bridge.

Mr. D. J. Whittemore, Chief Engineer of the Chicago, Milwaukee & St. Paul systems, says that on his roads* "everything not covered with earth, except cattle-guards, be the span ten or four hundred feet, is called a bridge. Everything covered with earth is called a culvert. Wherever we are far removed from suitable quarries, we build a wooden culvert in preference to a pile-bridge, if we can get six inches of filling over it. These culverts are built of roughly-squared logs, and are large enough to draw an iron pipe through them of sufficient diameter to take the water. We do this because we believe that we lessen the liability to accident, and that the culvert can be maintained, after decay has begun, much longer than a piled bridge with stringers to carry the track. Had we good quarries along our line, stone would be cheaper [in maintenance and final cost, but not in first cost.—F.]. Many thousands of dollars have been spent by this company in building masonry that, after twenty to twenty-five years, shows such signs of disintegration that we confine masonry work now only to stone that we can procure from certain quarries known to be good."

Mr. Whittemore is an engineer of great experience, skill, and judgment, and there is food for much reflection in these words of his. First, that it is better to use temporary wooden structures, to be afterward renewed in good stone, rather than to build of the stone of the locality, unless first-class. Second, that a structure covered with earth is much safer than an open bridge, which, if short and apparently insignificant, may be, through neglect, a most serious point of danger, as was shown in the dreadful accident of last year† on the Toledo, Peoria & Western Road in Illinois, where one hundred and fifty persons were killed and wounded, and by the equally avoidable accident on the Florida & Savannah line in March, 1888. Had these little trestles been changed to culverts covered with earth, many valuable lives would not have been lost.

* T. C. Clarke, in *Scribner's Magazine* for June, 1888, p. 657.   † 1887.

PART II.

STANDARD TRESTLE PLANS.

NOTE.—As the quantity of much of the material in a trestle varies with the height and of all of it with the length, it was considered better to merely give a list of dimensions of the different parts rather than a bill of material for some special height and length in the descriptions of the following examples of construction.

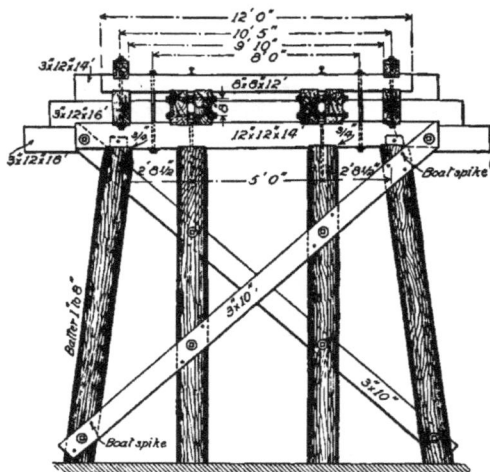

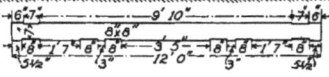

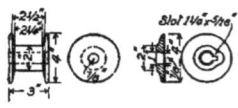

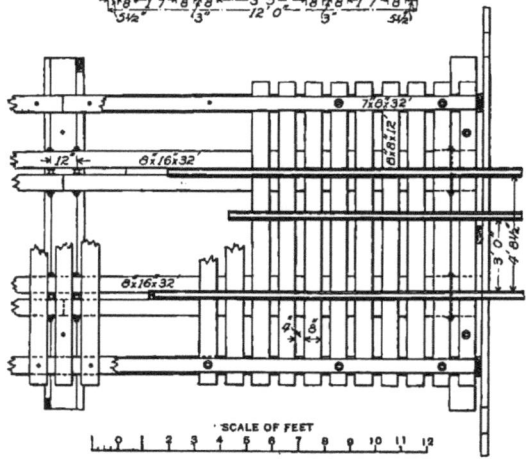

PLATE I.—STANDARD PILE-TRESTLE, DENVER & RIO GRANDE RAILROAD.

86

PART II.

STANDARD TRESTLE PLANS.

SECTION I.

PILE TRESTLES.

STANDARD PILE-TRESTLE, DENVER & RIO GRANDE RAILROAD.—PLATE I.

Dimensions of Timber.

Floor-System : Guard-rails, 7 in. × 8 in. × 32 ft., notched 1 in.

      Ties, 8 in. × 8 in. × 12 ft., notched ½ in. for both guard-rails and stringers, as shown in detail.

      Track-stringers, 8 in. × 16 in. × 32 ft., notched ¾ in. over caps.

      Jack-stringers, 8 in. × 16 in. × 32 ft., notched ¾ in. over caps.

Bent : Cap, 12 in. × 12 in. × 14 ft., notched ¾ in. over piles.

      Sway-braces, 3 in. × 10 in.

      Piles, 14 in. diameter at top.

Bank-bent : Dump-boards, 3 in. × 12 in. × 14 ft. ; 3 in. × 12 in. × 16 ft. ; 3 in. × 12 in. × 18 ft.

      Battens, 3 in. × 10 in. × 3 ft.

Dimensions of Iron Details.

Bolts : ¾ in. × 33 in. ; guard-rail to ties and jack-stringers.

      ¾ in. × 38 in. ; ties to caps.

      ¾ in. × 22 in. ; stringer-joints ; packing-bolts.

      ¾ in. × 18 in. ; sway-braces to posts.

Drift-bolts : ¾ in. × 22 in. ; caps to piles.

Boat-spikes : ½ in. × 8 in. ; sway-braces to posts.

Cast washers : ⅝ in. × 4 in. ; under heads and nuts of ¾-in. bolts.

Cast separators : 3 in. × 4 in. ; between stringer-pieces for ¾-in. bolts.

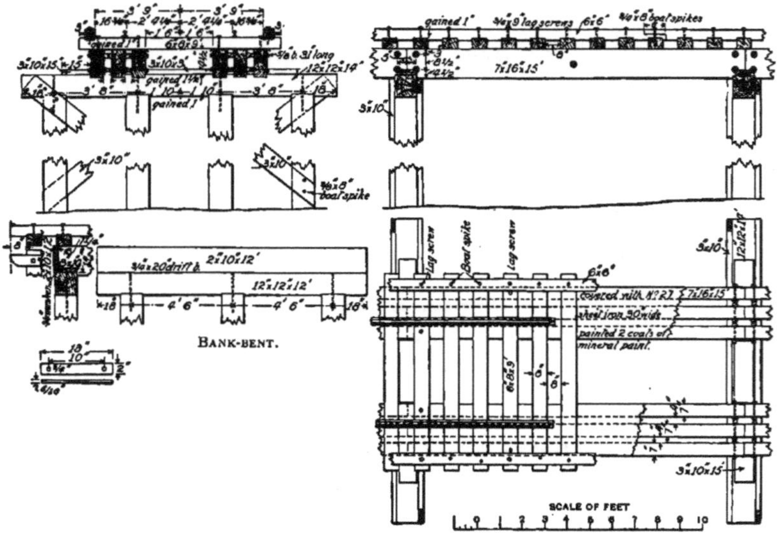

BANK-BENT.

SCALE OF FEET

PLATE II.—STANDARD PILE-TRESTLE, TOLEDO, ST. LOUIS & KANSAS CITY RAILROAD.

88

STANDARD PILE-TRESTLE, TOLEDO, ST. LOUIS & KANSAS CITY RAILROAD.—PLATE II.

Dimensions of Timber.

Floor-system : Guard-rails, 6 in. × 6 in., notched 1 in. over ties.

Ties, 6 in. × 8 in. × 9 ft.

Stringers, 7 in. × 16 in. × 15 ft., notched 1½ in. over caps.

Brace-blocks, 3 in. × 10 in. × 15 in. ; 3 in. × 10 in. × 3 ft.

Bents : Caps, 12 in. × 12 in. × 14 ft., notched 1 in. over piles.

Sway-braces, 3 in. × 10 in.

Piles, 4.

Bank-bent : Dump-board, 2 in. × 10 in. × 12 ft.

Cap, 12 in. × 12 in. × 12 ft.

Piles, 3.

Dimensions of Iron Details.

Bolts : ⅝ in. × 31 in. ; stringer-joints, packing-bolts.

Lag-screws : ¾ in. × 9 in. ; $\begin{cases} \text{guard-rails to ties.} \\ \text{ties to stringers.} \end{cases}$

Boat-spikes : ⅝ in. × 8 in. ; $\begin{cases} \text{guard-rails to ties.} \\ \text{sway-braces to posts.} \end{cases}$

Drift-bolts : ¾ in. × 20 in. ; caps to piles.

Wrought strap : 13 in. × 2 in. × $\frac{1}{16}$ in. ; stringer-joints.

Cast separators : 4 in. thick ; between stringers.

Cast washers : Under head and nut of each bolt.

Sheet-iron : No. 27, 30 in. wide ; to cover stringers.

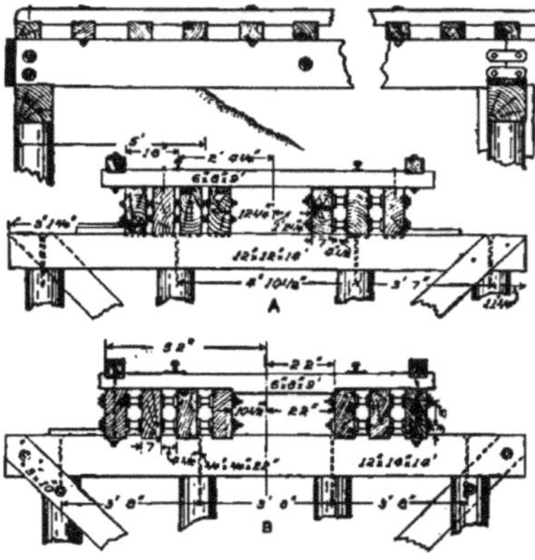

FIG. 1.—DETAILS OF FLOOR-SYSTEMS.

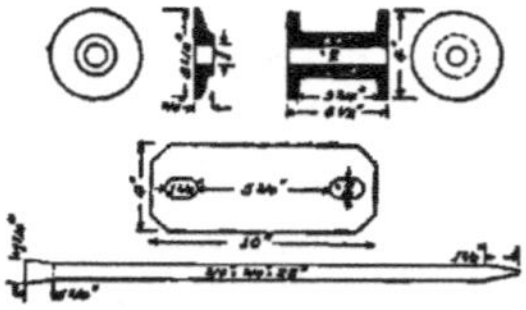

FIG. 3.—IRON DETAILS.

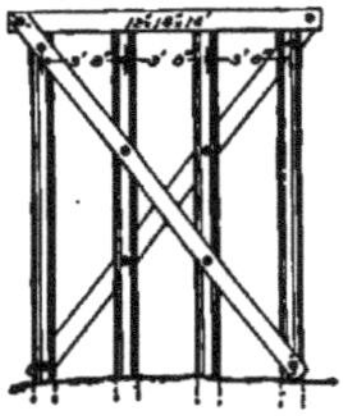

A, UP TO 18 FT. HIGH.

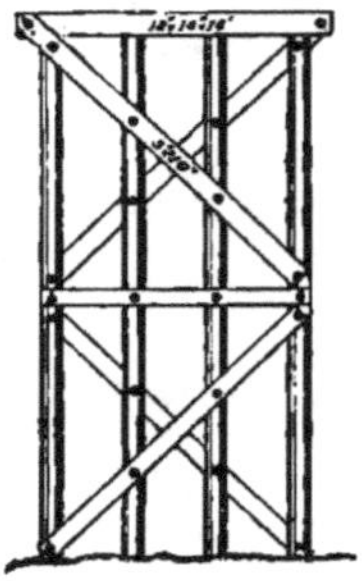

B, 19 TO 26 FT. HIGH.
FIG. 2.—PILE-BENTS.

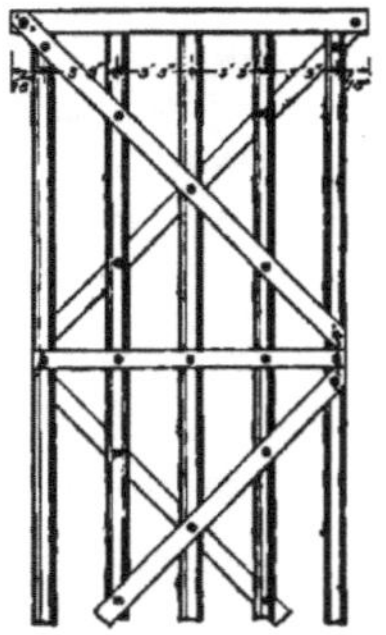

C, 27 TO 32 FT. HIGH.

PLATE III.—STANDARD PILE-TRESTLES, ATLANTIC & PACIFIC RAILROAD.

90

STANDARD PILE-TRESTLE, ATLANTIC & PACIFIC RAILROAD.—PLATE III.

(See also Plate XXXI.)

Dimensions of Timbers.

Floor-systems—Fig. 1, A: Guard-rails, 6 in. × 6 in. × 16 feet.

                    Ties, 6 in. × 8 in. × 9 ft.

                    Stringers, 7 in. × 16 in. × 15 ft.

                    Brace-blocks, 2 in. × 10 in. × 18 in.

           Fig. 1, B: same as above.

Bents: Caps, 12 in. × 14 in. × 14 ft.

       All sway-braces, 3 in. × 10 in.

       Piles, 12 in. diameter.

Dimensions of Iron Details.

Floor-system—Fig. 1, B; Four-piece stringer:

                Bolts, ¾ in. × 46 in.; stringer-joints.

                Packing-bolts, ¾ in. × 30 in.; guard-rails to stringers.

                Splice-plates: ⅜ in. × 4 in. × 10 in.; stringer-joints.

                Cast separators: 4 in. × 4½ in.; between stringer-pieces; ¾-in.

                    bolts.

                Cast washers: ¾ in. × 4½ in.; under head and nut of each bolt.

                Spikes: Boat, ⅝ in. × ⅝ in. × 7 in.;

                    Cut 20-penny.

            Three-piece stringer: Bolts, ¾ in.; × 34 in.; stringer-joints; pack-

                    ing-bolts.

                ¾ in. × 30 in.; guard-rails to stringers.

                Splice-plates, as above.

                Cast separators, as above.

                Cast washers, as above.

                Spikes: Boat, as above. Cut, as above.

Bents: Bolts, ¾ in. × 20 in.; sway-braces to piles.

       Drift-bolts: ¾ in. × ¾ in. × 22 in.; caps to piles.

       Boat-spikes: ⅝ in. × ⅝ in. × 7 in.; sway-braces to piles.

       Cast washers: ¾ in. × 4½ in.; under head and nut of each bolt.

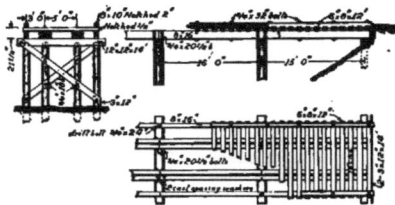

     Fig. 2.—Details of Separator and Cast Washers

PLATE IV.—STANDARD PILE-TRESTLE, CHICAGO & WEST MICHIGAN RAILWAY.

Standard Pile-trestle, Chicago & West Michigan Railway.—Plate IV.

(See also Plate X.)

Dimensions of Timbers.

Floor-system: Guard-rails, 8 in. × 10 in., notched 2 in. over ties.
Ties, 6 in. × 8 in. × 12 ft., notched ½ in. over stringers.
Stringers, 6 in. × 16 in. × 24 ft.

Bent: Cap, 12 in. × 12 in. × 14 ft.
Sway-braces, 3 in. × 12. in.
Piles, 12 in. diameter.

Bank-bent: Dump-plank, 3 in. × 12 in. × 16 ft.

Dimensions of Iron Details.

Bolts: ¾ in. × 32 in.; guard-rails to stringers.
¾ in. × 16½ in.; stringer-joints; packing-bolts.
¾ in. × 18 in.; sway-braces to piles.

Drift-bolts: ¾ in. × 24 in.; stringers to caps.

Boat-spikes: ⅝ in. × 7 in.; sway-braces to piles.

Cast washers: ¾ in. × in. 3½ in.; under head and nut of each bolt.
Other dimensions as per following table:

Bolt.	A	B	C	D	E	F	G	Weight.
⅝″—¾″	¾″	⅞″	1″	3½″	2″	½″	5/16″	1¼ lbs.
⅞″—1″	1″	1⅛″	1¼″	5″	2½″	⅝″	⅜″	
1⅛″—1¼″	1¼″	1⅜″	1½″	6″	3″	¾″	7/16″	
1⅜″—1½″	1½″	1⅝″	1¾	7¼″	3½	⅞″	½″	
1⅝″—1¾″	1¾″	1⅞″	2″	8½″	4½	1″	⅝″	
1⅞″—2″	2″	2⅛″	2¼″	9½″	5″	1⅛″	¾″	

Cast separators: 3 in. × 2 in. thick; between stringer-pieces.

These trestles are built with spans of 12 ft., 14 ft., and 16 ft.

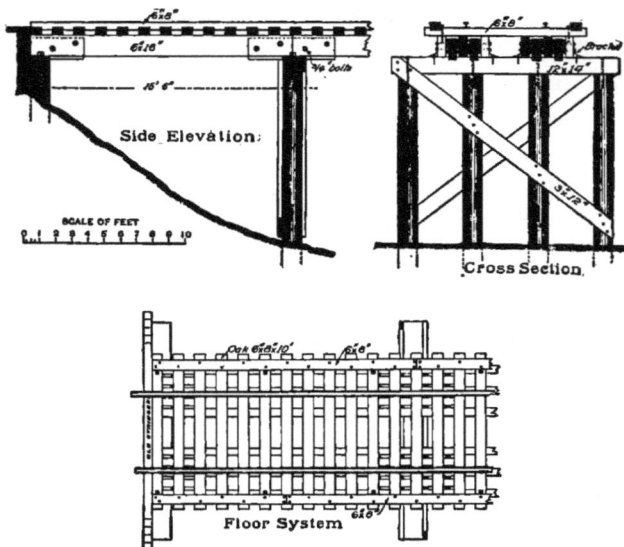

Side Elevation.

SCALE OF FEET

Cross Section.

Floor System

PLATE V.—STANDARD PILE-TRESTLE, MINNEAPOLIS & ST. LOUIS RAILWAY.

STANDARD PILE-TRESTLE, MINNEAPOLIS & ST. LOUIS RAILWAY.—PLATE V.

(See also Plate XXII.)

Dimensions of Timbers.

Floor-system : Guard-rail, 6 in. × 8 in., notched 2 in. over ties.

        Ties, 6 in. × 8 in. × 10 ft., white oak.

        Stringers, 6 in. × 16 in. × 15 ft. 6 in.

        Packing-block, 6 in. × 16 in. × 5 ft. 4 in., notched 2 in. over caps.

Bent : Cap, 12 in. × 14 in. × 14 ft., laid flat.

      Sway-braces, 3 in. × 12 in.

      Piles, not less than 11 in. diameter.

Bank-bent : Dump-plank, old stringers.

Dimensions of Iron Details.

Bolts : $\frac{3}{4}$ in. × 41 in.; stringer-joints ; packing-bolts.

Lag-screws : Stringer-brackets to caps.

Spikes : Boat, $\frac{1}{2}$ in. × 8 in.; guard-rails to ties.

Drift-bolts : $\frac{3}{4}$ in. × 12 in.; ties to stringers.

          $\frac{3}{4}$ in. × 22 in.; caps to piles.

Cast separators : 2 in. wide ; between packing-blocks and stringers.

Cast washers : Under head and nut of each bolt.

Cast brackets : Stringers to caps.

For arguments in favor of and description of this trestle, see *Railroad Gazette*, April 17, 1891.

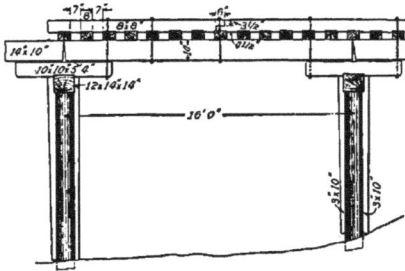

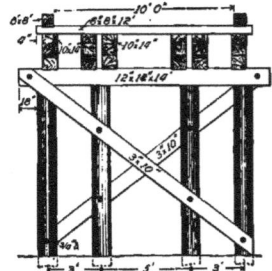

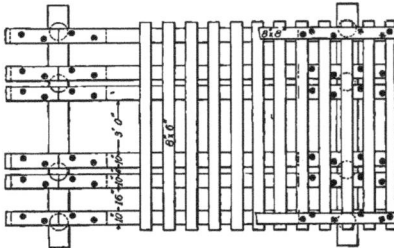

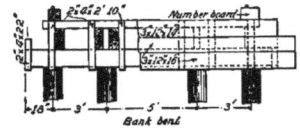

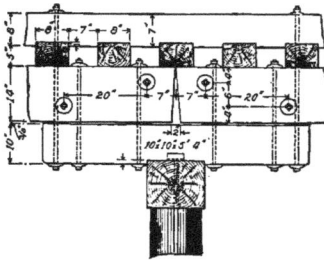

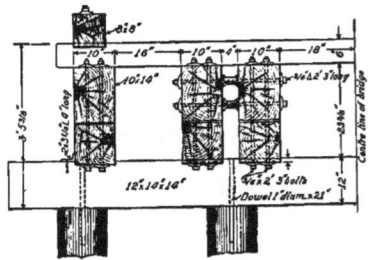

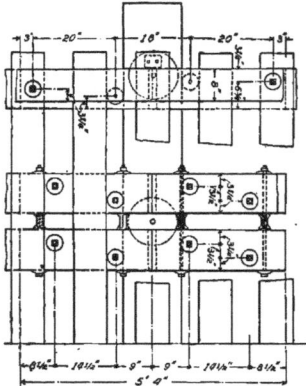

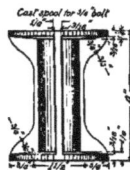

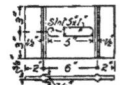

Corbel-stringer
Separator.

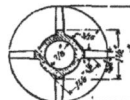

Cast-iron Spool.

Angle-lug.

IRON DETAILS.

DETAILS OF STRINGER-JOINT.

PLATE VI.—STANDARD PILE-TRESTLE, CHICAGO & NORTHWESTERN RAILWAY.

STANDARD PILE-TRESTLE, CHICAGO & NORTHWESTERN RAILWAY.—PLATE VI.

Dimensions of Timbers.

Floor-system : Guard-rails, 8 in. × 8 in. × 16 ft., notched 1 in. over ties.

　　　　　　Ties, 6 in. × 8 in. × 12 ft., white oak.

　　　　　　Track-stringers, 10 in. × 14 in. × 16 ft.

　　　　　　Jack-stringers : 10 in. × 14 in. × 16 ft.

　　　　　　Corbels, 10 in. × 10 in. × 5 ft. 4 in., notched 1 in. over cap, and used only
　　　　　　　　on bridges of two spans or over.

Bents : Cap, 12 in. × 14 in. × 14 ft.

　　　　Sway-braces, 3 in. × 10 in.

　　　　Piles, 12 in. diameter.

Bank-bent : Dump-plank, $\begin{cases} 3 \text{ in.} \times 10 \text{ in.} \times 16 \text{ ft.;} \\ 3 \text{ in.} \times 12 \text{ in.} \times 14 \text{ ft.;} \\ 3 \text{ in.} \times 12 \text{ in.} \times 16 \text{ ft.} \end{cases}$

　　　　Battens, $\begin{cases} 2 \text{ in.} \times 4 \text{ in.} \times 34 \text{ in.;} \\ 2 \text{ in.} \times 4 \text{ in.} \times 22 \text{ in.} \end{cases}$

Number-boards : 1½ in. × 8 in. × 12 in.

Dimensions of Iron Details.

Bolts : ¾ in. × 2 ft. 3 in.; stringer-joints ; packing-bolts ; also stringers to corbels.

　　　¾ in. × 2 ft. 5¾ in.; guard-rails to stringers.

　　　¾ in. × 3 ft. 4½ in.; guard-rails to corbels.

　　　¾ in. × 19¾ in.; sway-braces to caps and piles.

Dowels : 1 in. × 21 in.; caps to piles.

Spikes : Boat, $\frac{5}{16}$ in. × 5 in.

　　　　Cut 30-penny.

Cast separators : 3 in. × 4 in.; as per detail drawing ; between stringer-pieces.

　　　　　　　　6 in. × 10 in. ; as per detail drawing ; between stringers and corbels.

Cast washers : Under head and nut of each bolt.

Angle-iron lugs : 2 in. × 3¼ in. L × 4 in. long ; hold stringers in place.

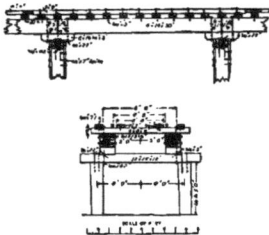

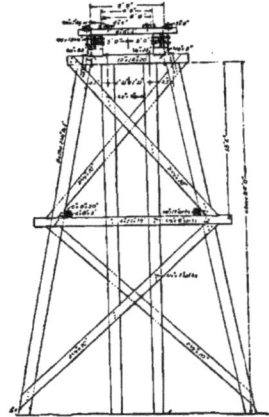

PLATE VII.—STANDARD PILE-TRESTLE, LOUISVILLE & NASHVILLE RAILROAD.

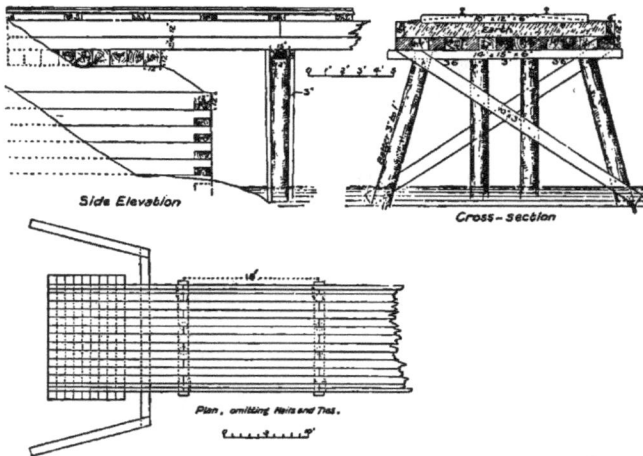

Side Elevation

Cross-section

Plan, omitting Rails and Ties.

PLATE VIII.—PILE-TRESTLE WITH EARTH ROADBED, LOUISVILLE & NASHVILLE RAILROAD.

98

STANDARD PILE-TRESTLE, LOUISVILLE & NASHVILLE RAILROAD.—PLATE VII.
(See also Plates VIII and XVII.)

Dimensions of Timbers.

Floor System : Guard-rails, outside, 5 in. × 9 in., notched 1 in. over ties.

inside, 4 in. × 9 in., not notched.

Ties, 6 in. × 8 in. × 9 ft.

Stringers, 8 in. × 16 in. × 30 ft.

Corbels, 8 in. × 16¼ in. × 3 ft., notched 1 in. over caps.

Bents : Caps, 12 in. × 14 in. × 12 feet.

Sway-braces, diagonal, 2½ in. × 10 in.

horizontal, 4 in. × 10 in.

Piles, 12 in. diameter.

Girts : 6 in. × 8 in. × 30 ft.

Splice-blocks : 2 in. × 8 in. × 3 ft.

Dimensions of Iron Details.

Bolts : ¾ in. × 27 in.; stringers to corbels.

¾ in. × 19¼ in.; stringer-joints : packing-bolts.

¾ in. × 42 in.; floor-system to caps.

Drift-bolts : ⅞ in. × 22 in.; caps to piles.

Dowels : ¾ in. × 5 in.; ties to stringers.

Spikes : ⅝ in. × 9 in.; corbels to caps.

½ in. × 8 in.; horizontal sway-braces to piles.

⅜ in. × 7 in. ; diagonal sway-braces to piles.

½ in. × 12 in.; girts to piles.

Lag-screws : ¾ in. × 7 in.; guard-rails to ties.

Cast washers : Under head and nut of each bolt.

Cast separators : Between stringer-pieces.

PILE-TRESTLE WITH EARTH ROADBED, LOUISVILLE & NASHVILLE RAILROAD.—
PLATE VIII. (See also Plates VII and XVII.)

Dimensions of Timbers.

Ties, 6 in. × 12 in. × 10 ft.

Side-timbers, 6 in. × 12 in. × 32 ft.

Floor-timbers, 8½ in. × 12 in. × 32 ft.

Caps, 6 in. × 15 in. × 14 ft.

Sway-bracing, 3 in. × 10 in.

Piles, 12 in. diameter.

Revetment-timbers, 12 in. × 12 in.

All timber creosoted yellow pine, spiked together.   No bolts or mortise and tenon joints used.—*Eng. News*, Oct. 29, 1887.

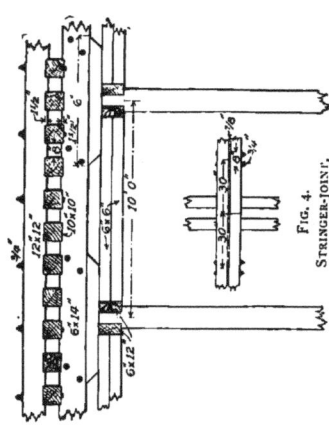

FIG. 4.
STRINGER-JOINT.

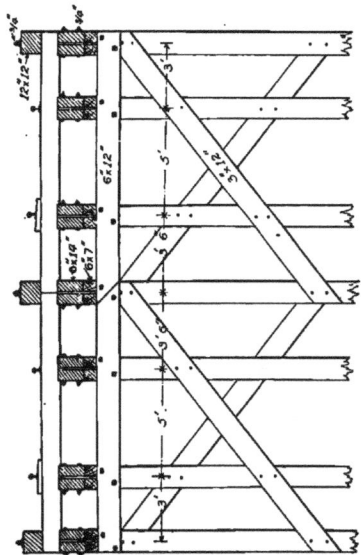

SCALE OF FEET

FIG. 1.—CROSS-SECTION.

FIG. 2.—SIDE ELEVATION.

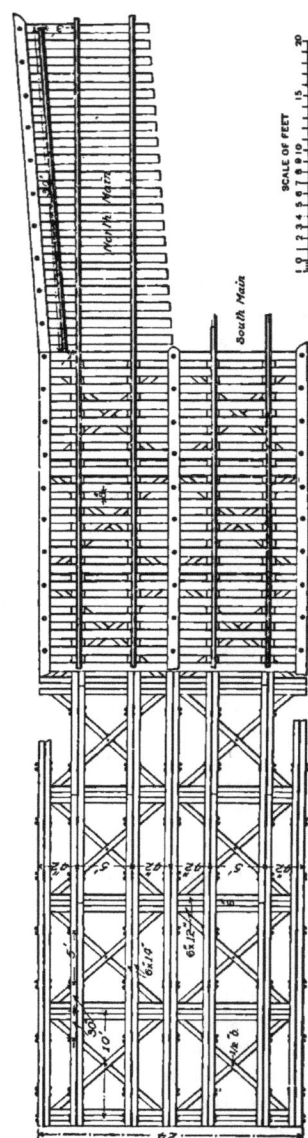

North Main

South Main

SCALE OF FEET

FIG. 3.—PLAN.

PLATE IX.—STANDARD PILE-TRESTLE, BOSTON & ALBANY RAILROAD—BRIDGE No. 35.

STANDARD PILE-TRESTLE, BOSTON & ALBANY RAILROAD.—PLATE IX.

Dimensions of Timbers.

Floor-system : Guard-rails, 12 in. × 12 in., notched $1\frac{1}{2}$ in. over ties.

Ties, 10 in. × 10 in. × 12 ft., notched $1\frac{1}{2}$ in. over stringers.

Stringers, 6 in. × 14 in. × 30 ft.

Corbels, 6 in. × 7 in. × 6 ft.

Bents : Caps, 6 in. × 12 in. × 12 ft. 6 in.

Sway-braces, 3 in. × 12 in.

Piles, 12 in. diameter.

Lateral braces : 6 in. × 6 in.

Dimensions of Iron Details.

Bolts : $\frac{3}{4}$ in. × $23\frac{1}{2}$ in.; guard-rails to ties.

$\frac{3}{4}$ in. × 16 in.; stringer-joints; packing-bolts.

$\frac{3}{4}$ in. × 21 in.; caps to piles.

$\frac{1}{2}$ in. × $13\frac{1}{2}$ in.; lateral brace intersections.

Spikes.

Cast separators.

Cast washers.

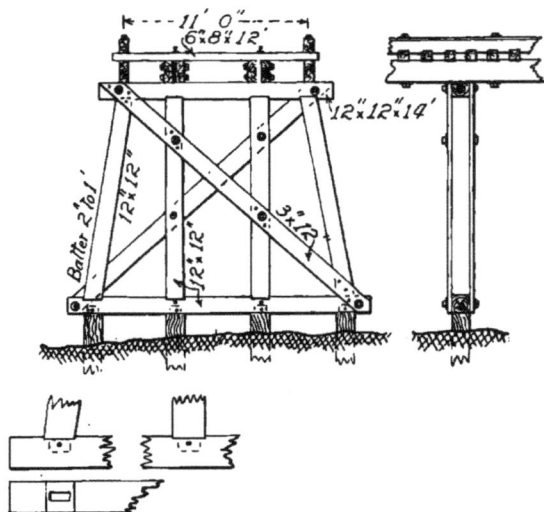

PLATE X.—STANDARD FRAMED TRESTLE, CHICAGO & WEST MICHIGAN RAILWAY.

SECTION II.

FRAMED TRESTLES.

STANDARD FRAMED TRESTLE, CHICAGO & WEST MICHIGAN RAILWAY.—PLATE X

(See also Plate IV.)

Dimensions of Timbers.

Floor-system : Guard-rails, 8 in. × 10 in., notched 2 in. over ties.

Ties, 6 in. × 8 in. × 12 ft., notched ½ in. over stringers.

Stringers, 8 in. × 16 in. × 32 ft.

Bent : Cap, 12 in. × 12 in. × 14 ft.

Plumb-posts, 12 in. × 12 in.

Batter-posts, 12 in. × 12 in. ; batter, 2 in. to 1 ft.

Sill, 12 in. × 12 in.

Sway-braces, 3 in. × 12 in.

Sub-sills, 12 in. × 12 in. × 6 ft

Dimensions of Iron Details.

Same as for Plate IV

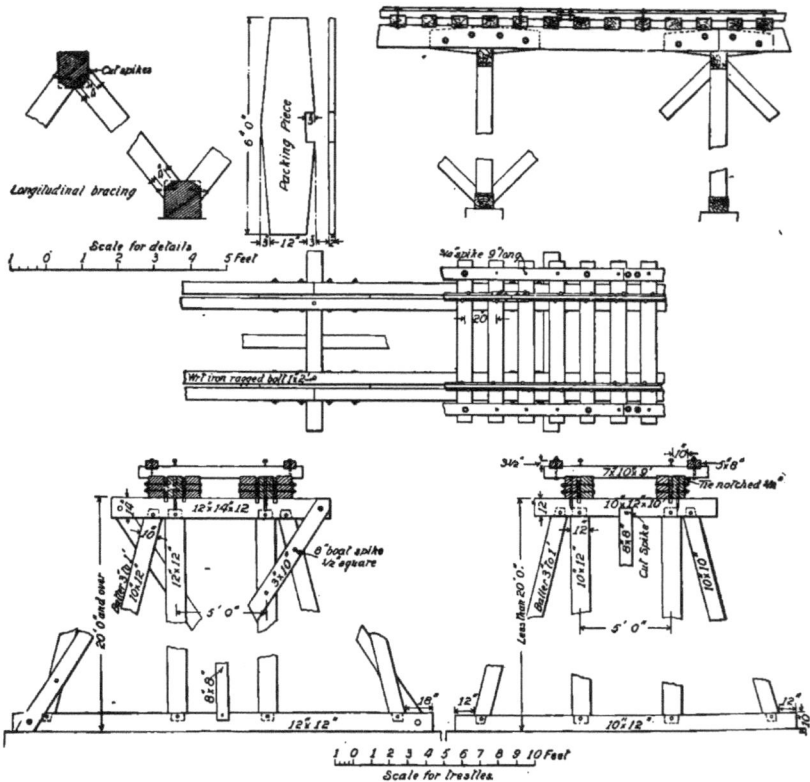

PLATE XI.—STANDARD FRAMED TRESTLES, PENNSYLVANIA RAILROAD.

STANDARD FRAMED TRESTLE, PENNSYLVANIA RAILROAD.—PLATE XI.

Dimensions of Timbers.

Floor-system : Guard-rails, 5 in. × 8 in., notched 1 in. over ties.

Ties, 7 in. × 10 in. × 9 ft., notched ½ in. to receive guard-rails, and ½ in. over stringers.

Stringers :

Clear Span.	Number of Pieces under each Rail.	Width of each Piece.	Depth of Stringers.
10 ft.	2	8 in.	15 in.
12 "	2	8 "	16 "
14 "	2	10 "	17 "
16 "	3 -	8 "	17 "

Packing-blocks, 2 in. × 18 × 6 ft.

Bents under 20 ft. : Cap, 10 in. × 12 in. × 10 ft.

Plumb-posts, 10 in. × 12 in.

Batter-posts, 10 in. × 10 in. ; batter, 3 in. to 1 ft.

Sill, 10 in. × 12 in.

Bents 20 ft. and over : Cap, 12 in. × 14 in. × 12 ft.

Plumb-posts, 12 in. × 12 in.

Batter-posts, 10 in. × 12 in., batter 3 in. to 1 ft.

Sill, 12 in. × 12 in.

Sway-bracing, 3 in. × 10 in.

Bracing : Longitudinal, 8 in. × 8 in.

Treenails : Locust, 1 in. diameter.

Dimensions of Iron Details.

Bolts : ¾ in. ×——; guard-rails to ties.

¾ in. ×——; guard-rail joints.

¾ in. ×——; stringer-joints ; packing-bolts.

All of above bolts have 2½-in. flat heads, with 2½-in. wrought washer under nuts.

⅞ in. ×——, sway-bracing to caps and sills ; 3-in. wrought-iron washers used.

Drift-bolts (ragged) : 1 in. × 24 in. ; stringers to caps.

Spikes : Boat, ¾ in. × 9 in. ; guard-rails to ties.

½ in. × 8 in. ; sway-braces to posts.

Cut ——×——, longitudinal braces to caps and sills.

Wrought washers : 2½ in. square for ¾-in. bolt.

3 in. round for ¾-in. bolt.

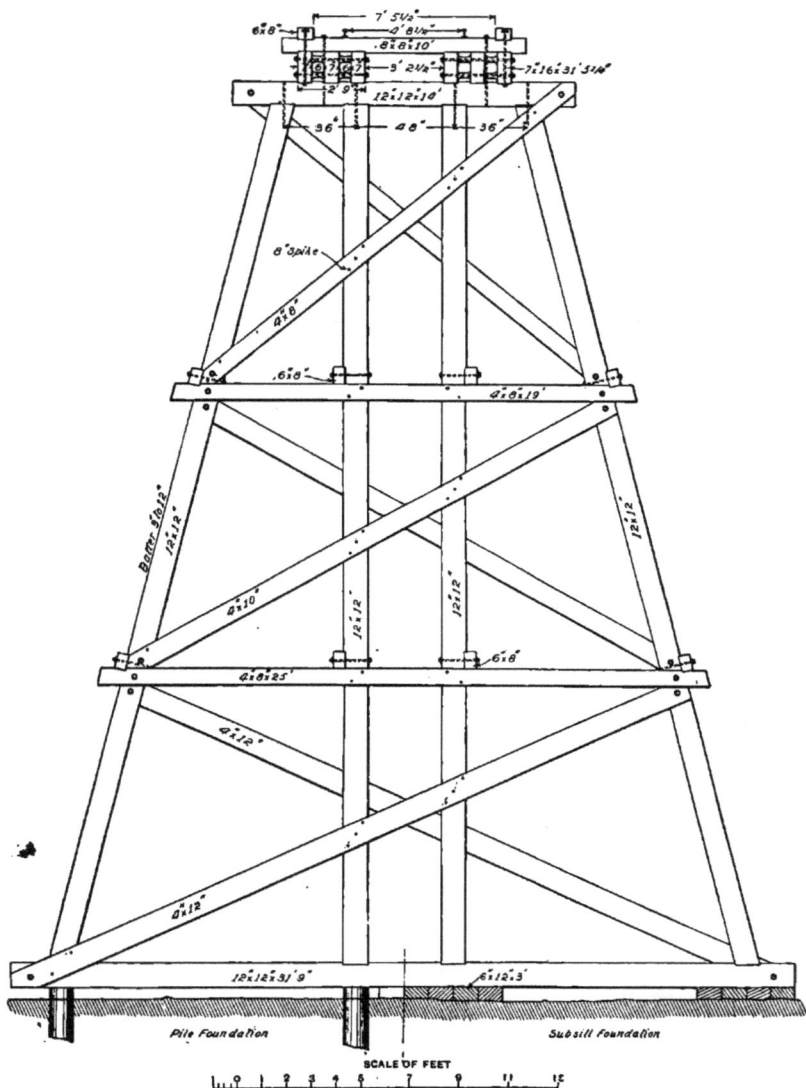

PLATE XII.—STANDARD FRAMED TRESTLE, SAN FRANCISCO & NORTH PACIFIC RAILROAD.

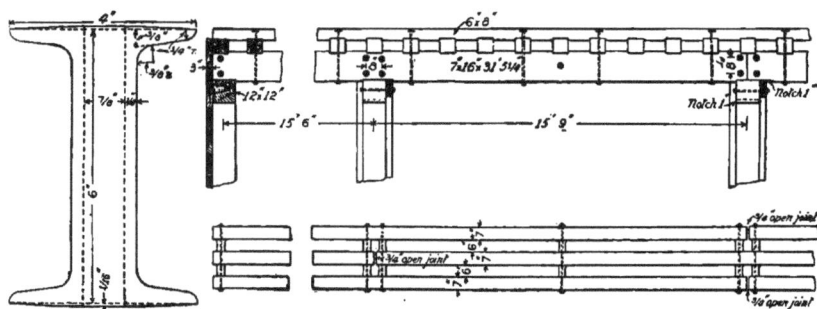

DETAILS OF FLOOR-SYSTEM.

PLATE XIII.—STANDARD FRAMED TRESTLE, SAN FRANCISCO & NORTH PACIFIC RAILROAD.

STANDARD FRAMED TRESTLE, SAN FRANCISCO & NORTH PACIFIC RAILROAD.—
PLATES XII AND XIII.

Dimensions of Timbers.

Floor-system : Guard-rails, 6 in. × 8 in., notched over ties.

Ties, 8 in. × 8 in. × 10 ft., notched over stringers.

Stringers, 7 in. × 16 in. × 31 ft. 5¼ in., notched 1 in. over caps.

Bents : Caps, 12 in. × 12 in. × 14 ft.

Plumb-posts, 12 in. × 12 in.

Batter-posts, 12 in. × 12 in.; batter, 3 in. to 1 ft.

Sill, 12 in. × 12 in.

Sway-braces : Horizontal, 4 in. × 8 in.

Diagonal, 4 in. × 8 in., 4 in. × 10 in., 4 in. × 12 in.

Longitudinal bracing : Girts, 6 in. × 8 in.

Sub-sills : 6 in. × 12 in. × 3 ft.

Bank-bent : Dump-boards, 3 in. × 12 in. × 14 ft.

Dimensions of Iron Details.

Bolts : ¾ in. × 37 in.; floor-system to cap.

⅝ in. × 36 in.; stringer joints; packing bolts.

⅝ in. × 28¼ in.; guard-rails to ties and stringers.

⅝ in. × 21½ in.; horizontal sway-braces to posts.

⅝ in. × 18½ in.; longitudinal braces to posts.

⅝ in. × 17½ in.; diagonal sway-braces to posts, etc.

Drift-bolts : —— × ——; cap to posts.

—— × ——; sill to piles.

Spikes : 8 in.; sway-braces to posts, etc.

Cast separators : 4 in. × 6 in. thick; between stringer-pieces.

Cast washers for ⅝-in. and ¾-in. bolts.

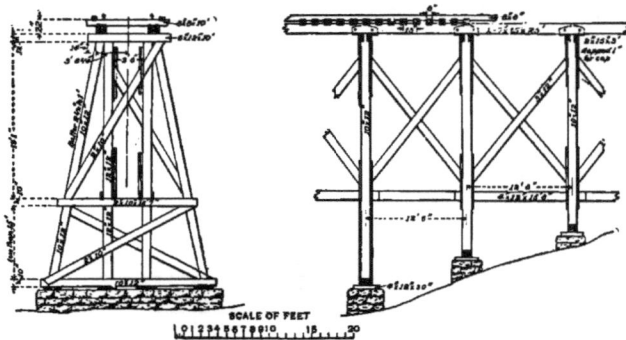

FIG. 1.—CROSS-SECTION.   FIG. 2.—ELEVATION.

GENERAL PLAN SINGLE-DECK TRESTLES.

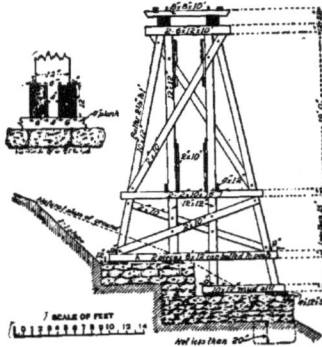

SCALE OF FEET

FIG. 3.—PLAN FOR BREAKING SILLS AND STEPPING FOOTINGS ON STEEP SLOPES.

PLATE XIV.—STANDARD TRESTLES, NORFOLK & WESTERN RAILROAD.

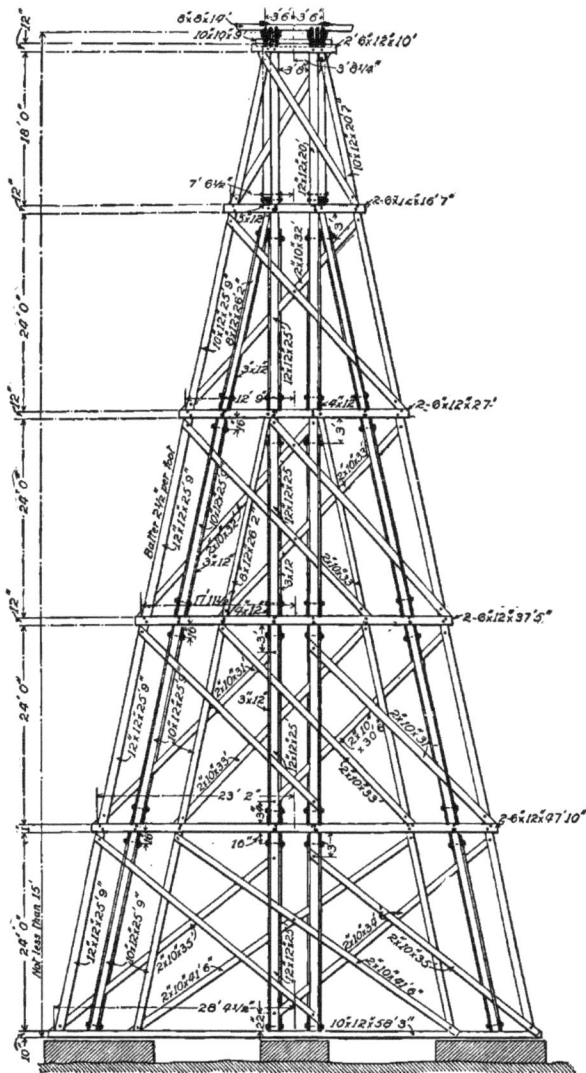

SCALE OF FEET
0 1 2 3 4 5   10   15   20   5   30

FIG. 4.—CROSS-SECTION HIGH OR MULTIPLE STORY TRESTLE.

PLATE XV.—STANDARD TRESTLE, NORFOLK & WESTERN RAILROAD.

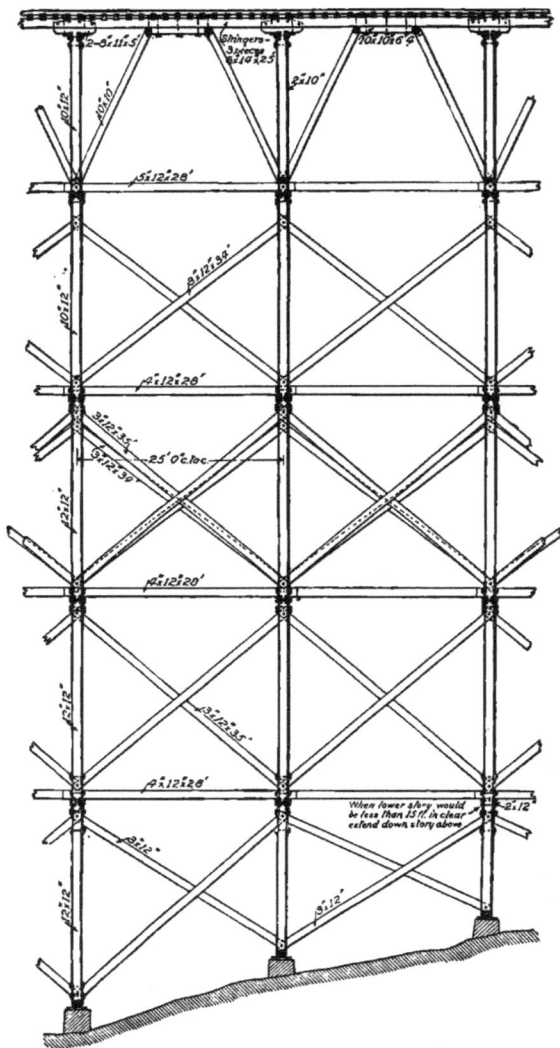

FIG. 5.—ELEVATION HIGH OR MULTIPLE STORY TRESTLE.

PLATE XVI.—STANDARD TRESTLE, NORFOLK & WESTERN RAILROAD.

STANDARD FRAMED TRESTLE, NORFOLK & WESTERN RAILROAD.—
PLATES XIV, XV, AND XVI.

PLATE XIV.

Dimensions of Timbers.

Floor-system : Guard-rails, 6 in. × 8 in., notched.

Ties, 8 in. × 8 in. × 10 ft., notched.

Stringers, 7 in. × 15 in. × 25 ft.

Packing-blocks, 2 in. × 15 in. × 3 ft., notched 1 in. over cap.

Bent : Cap, 6 in. × 12 in. × 10 ft.

Plumb-posts, 12 in. × 12 in.

Batter-posts, 10 in. × 12 in.; batter, 2½ in. to 1 ft.

Sill, 10 in. × 12 in.

Sway-bracing : Diagonal, 2 in. × 10 in.

Horizontal, 2 in. × 10 in.

Longitudinal bracing : Horizontal, 4 in. × 12 in. × 15 ft. 6 in.

Diagonal, 3 in. × 12 in.

Sub-sills : 4 in. × 12 in. × 2 ft. 6 in.

PLATES XV AND XVI.

Floor-system : Guard-rails, 6 in. × 8 in.

Ties, 8 in. × 8 in. × 14 ft.

Stringers, 6 in. × 14 in. × 25 ft.

Packing-blocks, 2 in. × 14 in. × 5 ft.

Corbels, 8 in. × 11 in. × 5 ft.

Bent : Cap, 6 in. × 12 in. × 10 ft.

Plumb-posts, 12 in. × 12 in.

Outside batter-posts, 10 in. × 12 in., and 12 in. × 12 in.

Inside batter posts, 8 in. × 12 in., and 10 in. × 12 in.

Sway-braces, 2 in. × 10 in.

Intermediate caps, 6 in. × 12 in.

Sill, 10 in. × 12 in.

Longitudinal bracing : Horizontal, 4 in. × 12 in. × 28 ft.

Diagonal, 3 in. × 12 in.

Knee-braces : Straining-beam, 10 in. × 10 in. × 9 ft.

Top chord, 10 in. × 10 in. × 6 ft. 4 in.

Bottom chord, 5 in. × 12 in. × 28 ft.

Diagonals, 10 in. × 10 in.

For method of elevating rails on curves, see Part I, Fig. 125.

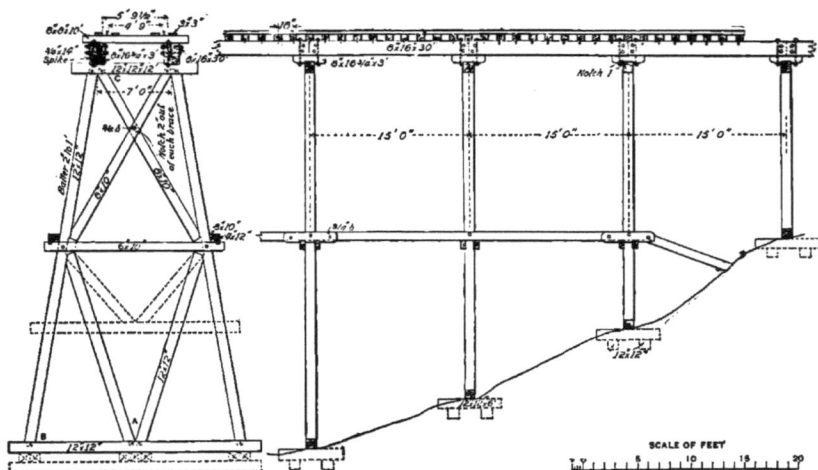

FIG. 1.—GENERAL PLANS.

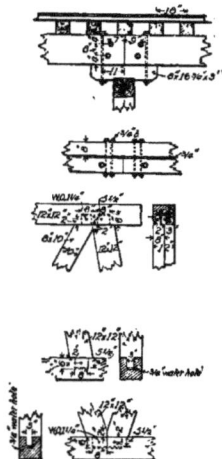

SCALE OF FEET

FIG. 3.—DETAIL OF JOINT OF
DIAGONAL POSTS.

FIG. 2.—DETAILS OF STRINGER
AND POST JOINTS.

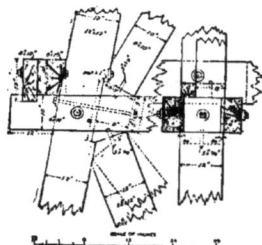

PLATE XVII.—STANDARD FRAMED TRESTLE, LOUISVILLE & NASHVILLE RAILROAD.

112

STANDARD FRAMED TRESTLE, LOUISVILLE & NASHVILLE RAILROAD.—PLATE XVII.
(See also Plates VII and VIII.)

Dimensions of Timbers.

Floor-system : Guard-rails, 3 in. × 9 in.
      Ties, 8 in. × 8 in. × 10 ft.
      Stringers, 8 in. × 16 in. × 30 ft.
      Corbels, 8 in. × 16¾ in. × 3 ft.
Bent : Cap, 12 in. × 12 in. × 12 ft.
   Batter-posts, 12 in. × 12 in.
   Diagonal posts, 8 in. × 10 in., notched 2 in. each at intersection.
   Intermediate cap, 6 in. × 10 in.
   Sill, 12 in. × 12 in.
Longitudinal braces : Horizontal, 8 in, × 10 in. × 30 ft.
       Splice-block, 4 in. × 12 in.
Sub-sills : 12 in. × 12 in. × 6 ft.

Dimensions of Iron Details.

Bolts : ¾ in. × 27 in.; stringers to corbels and intermediate cap to posts.
   ¾ in. × 20 in. ; stringer-joints ; packing-bolts.
   ¾ in. × 15 in. ; splice-block to girts.
   ¾ in. × 23 in. ; angle-block to posts.
Spikes : ⅝ in. × 14 in. ; corbels to caps.
Dowels : ⅝ in. × 5 in. ; diagonal posts to angle-blocks.
Cast washers : Under head and nut of each bolt.
Cast separators : ¾ in. × ——— ; between stringer-pieces.

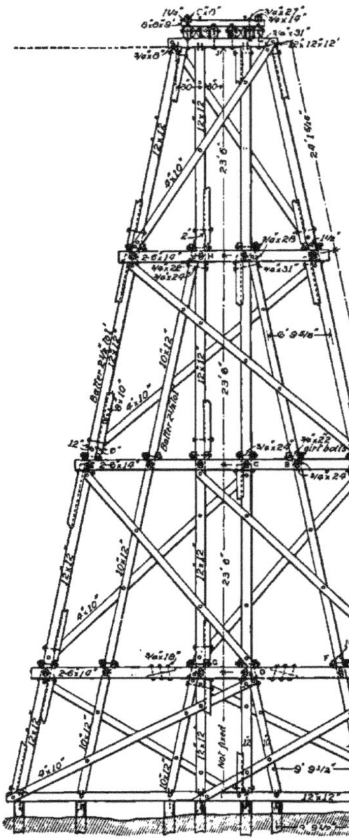

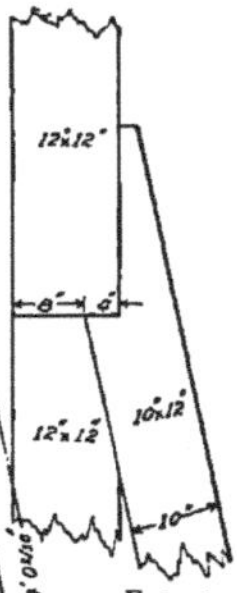

FIG. 4.
JOINT H.

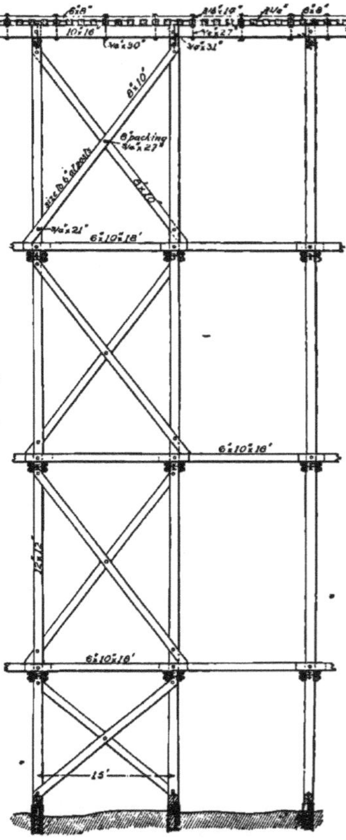

SCALE OF FEET

FIG. 1.—CROSS-SECTION.

FIG. 2.—ELEVATION.

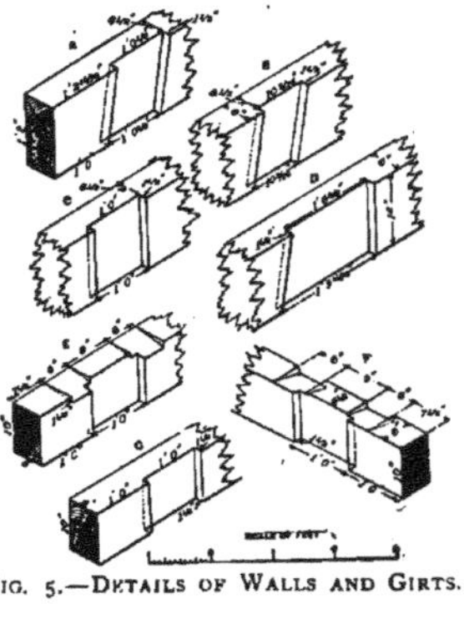

FIG. 3.—FLOOR-SYSTEM.

FIG. 5.—DETAILS OF WALLS AND GIRTS.

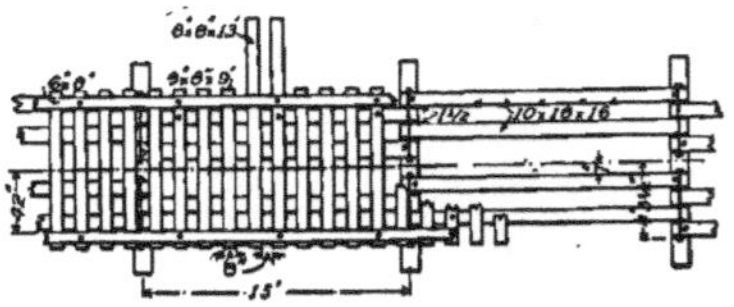

PLATE XVIII.—STANDARD FRAMED TRESTLE,
OREGON PACIFIC RAILROAD.

114

STANDARD FRAMED TRESTLE, OREGON PACIFIC RAILROAD.—PLATE XVIII.

Dimensions of Timbers.

Floor-system : Guard-rails, 6 in. × 8 in., notched 1¼ in.

Ties, 8 in. × 8 in. × 9 ft., not notched ; and two ties 13 ft. long for every fourth span projecting on alternate sides.

Stringers, 10 in. × 16 in. × 16 ft., not notched.

Bent : Caps, 12 in. × 12 in. × 12 ft.

Plumb-posts, 12 in. × 12 in., in 23 ft. 6 in. lengths.

Outside batter-posts, 12 in. × 12 in., in 24 ft. $0\tfrac{1}{16}$ in.* lengths.

Counter-posts or inside batter-posts, 10 in. × 12 in., in 24 ft. $0\tfrac{1}{16}$* in. lengths.

Intermediate caps or horizontal sway-bracing, 6 in. × 14 in.

Diagonal sway-bracing, 4 in. × 10 in.

Sill, 12 in. × 12 in.

Longitudinal bracing : Girts, 6 in. × 10 in. × 18 ft., notched 1½ in.

Diagonals, 8 in. × 10 in., sized to 6 in. at posts.

Packing-pieces, 8 in. thick at intersection of diagonals.

Dimensions of Iron Details.

Bolts : ¾ in. × 14 in. ; guard-rails to ties.

¾ in. × 30 in. ; through guard-rails, ties, and outside stringers.

¾ in. × 27 in. ; ties to stringers.

¾ in. × 31 in. ; stringers to caps.

¾ in. × 48 in. ; stringer-joints ; packing-bolts.

¾ in. × 18 in. ; diagonal sway-braces to posts.

¾ in. × 28 in. ; ⎫
¾ in. × 22 in. ; ⎪ bolt at joint H and its companion joint.
¾ in. × 24 in. ; ⎪
¾ in. × 31 in. ; ⎭

¾ in. × 24 in. ; intermediate caps to posts.

¾ in. × 18 in. ; sill-joint bolts.

¾ in. × 22 in. ; ⎱ girt-bolts.
¾ in. × 24 in. ; ⎰

¾ in. × 21 in. ; diagonal longitudinal braces to posts.

¾ in. × 27 in. ; intersection of above.

Dowels : ¾ in. × 8 in. ; cap and sill to posts ; post-joints.

Drift-bolts : —— × ——

Cast washers : Under heads and nuts of each bolt.

Cast separators : 1½ in. × —— ; between stringer-pieces.

---

* So in original blue print, but rather too close to work to in this size timber.

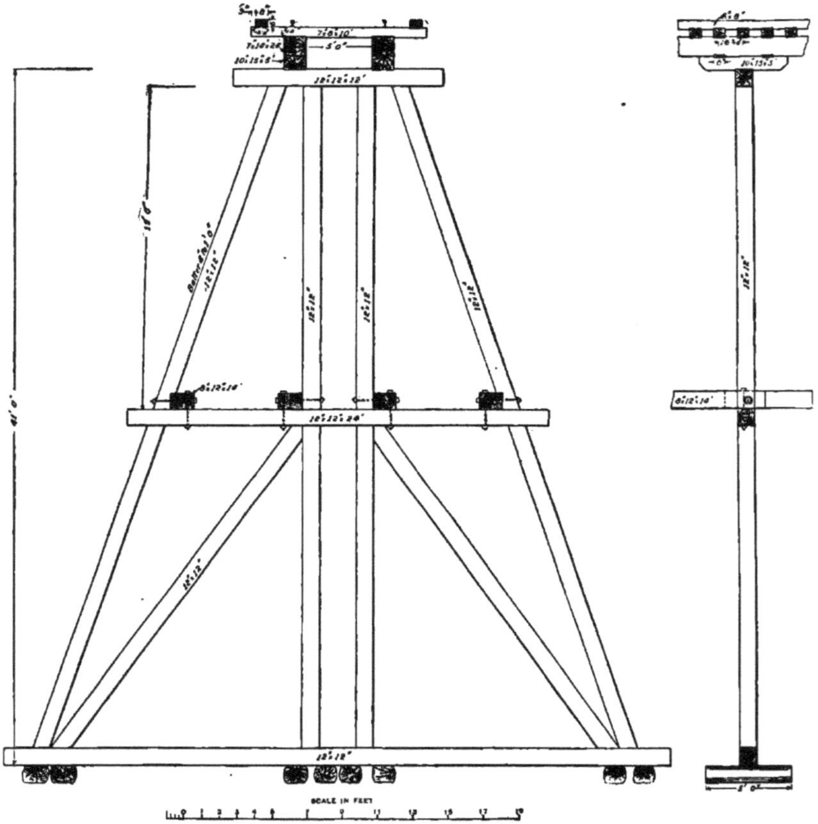

PLATE XIX.—FRAMED TRESTLE, OHIO CONNECTING RAILWAY.

FRAMED TRESTLE, OHIO CONNECTING RAILWAY.—PLATE XIX.

Dimensions of Timbers.

Floor-system : Guard-rails, 6 in. × 8 in., notched ¾ in. over ties.

Ties, 7 in. × 8 in. × 10 ft., notched ¾ in. over stringers.

Stringers, 7 in. × 14 in. × 24 ft.

Corbels, 10 in. × 15 in. × 5 ft., notched over caps.

Bents : Caps, 12 in. × 12 in. × 12 ft.

Plumb-posts, 12 in. × 12 in.

Batter-posts, 12 in. × 12 in.

Counter-posts, 12 in. × 12 in.

Intermediate caps, 12 in. × 12 in.

Sills, 12 in. × 12 in.

Longitudinal braces, 8 in. × 12 in. × 14 ft.

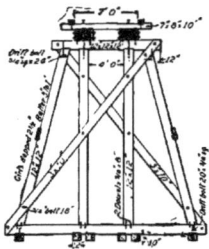

FIG. 1.—CROSS-SECTION.

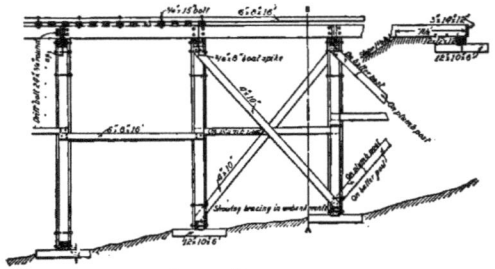

FIG. 2.—ELEVATION.

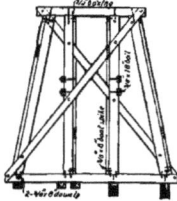

FIG. 5.—PACKING-WASHER.

FIG. 4.—SECTION A B, SHOWING BRACING IN EMBANKMENT.

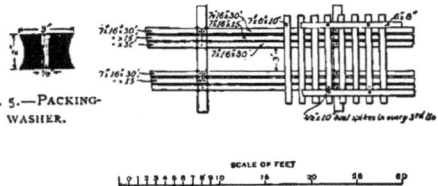

FIG. 3.—PLAN.

SCALE OF FEET

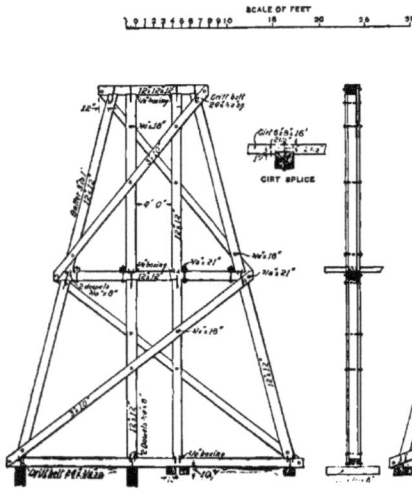

FIG. 6.—DOUBLE-DECK TRESTLE.

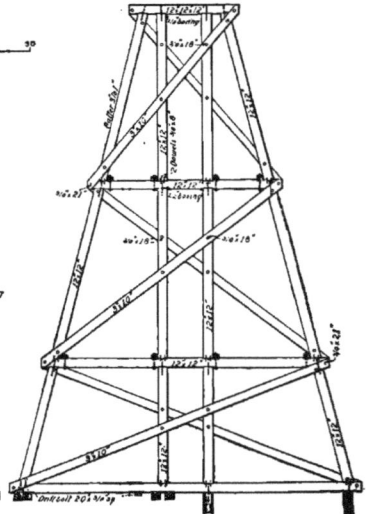

FIG. 7.—TRIPLE-DECK TRESTLE.

PLATE XX.—PRESENT STANDARD TRESTLE, CHARLESTON, CINCINNATI & CHICAGO RAILROAD.

118

STANDARD FRAMED TRESTLE, CHARLESTON, CINCINNATI & CHICAGO RAILROAD.—
PLATES XX AND XXI.

PLATE XX.

Dimensions of Timbers.

Floor-system : Guard-rails, 6 in. × 8 in. × 16 ft.
        Ties, 7 in. × 8 in. × 10 ft.
        Stringers, 7 in. × 16 in. × 30 ft. and 15 ft.
Bent · Cap, 12 in. × 12 in. × 12 ft.
    Plumb-posts, 12 in. × 12 in.
    Batter-posts, 12 in. × 12 in.
    Sway-bracing, 3 in. × 10 in.
    Intermediate cap, 12 in. × 12 in.
    Sill, 12 in. × 12 in.
Longitudinal bracing : Horizontal, 6 in. × 8 in. × 16 ft.
            Diagonal, 4 in. × 10 in.
Sub-sills, 10 in. × 12 in. × 6 ft.

Dimensions of Iron Details.

Bolts : $\frac{3}{4}$ in. × 15 in.; guard-rails to ties.
    $\frac{3}{4}$ in. × 18 in.; sway-braces to posts.
    $\frac{3}{4}$ in. × 28 in.: stringer-joints ; packing-bolts.
Drift-bolts : $\frac{3}{4}$ in. × $\frac{3}{4}$ in. × 20 in.; sills to sub-sills.
        $\frac{3}{4}$ in. × $\frac{3}{4}$ in. × 24 in.; stringers to caps ; caps to posts.
Dowels : $\frac{3}{4}$ in. × 8 in. ; posts to sills.
Spikes, boat : $\frac{5}{8}$ in. × 8 in. ; girts to posts.
        $\frac{1}{2}$ in. × 10 in.; ties to stringers.
Cast separators : 2 in. × 3 in.; between stringer-pieces.
Cast washers : —— × 3 in. ; under head and nut of each bolt.

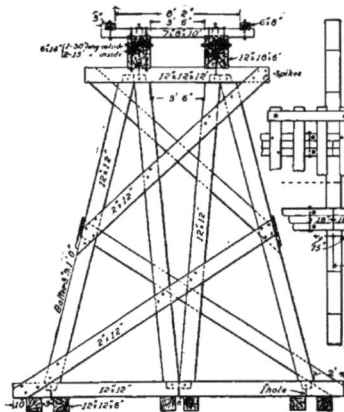

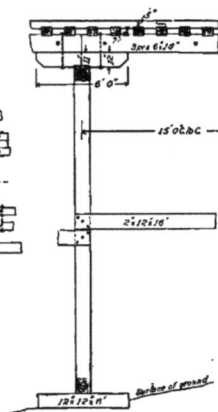

FIG. 1.—CROSS-SECTION.　FIG. 3.—PLAN.　FIG. 2.—ELEVATION.

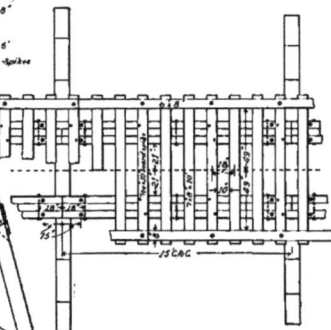

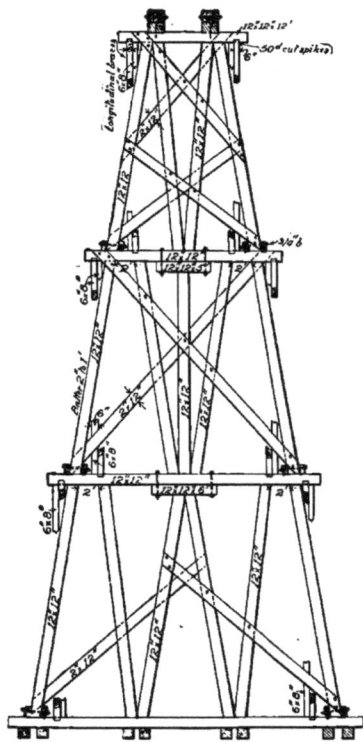

SINGLE-DECK TRESTLE.

SCALE OF FEET

FIG. 4.—CROSS-SECTION.　FIG. 5.—ELEVATION.

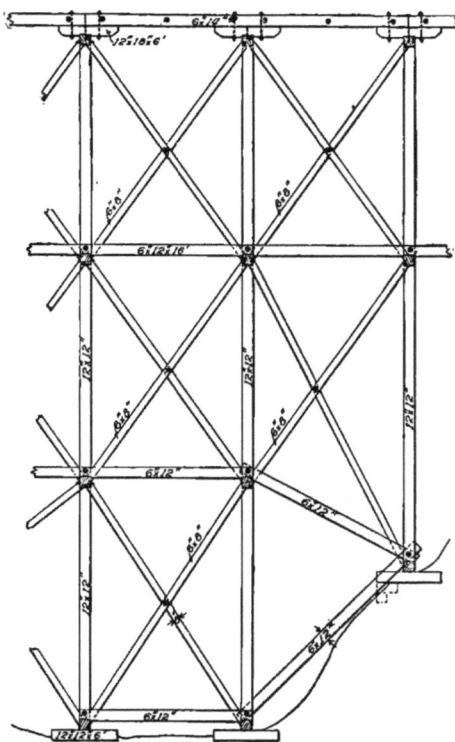

TRIPLE-DECK TRESTLE.

PLATE XXI.—FORMER STANDARD TRESTLE, CHARLESTON. CINCINNATI & CHICAGO RAILROAD

PLATE XXI.

Dimensions of Timbers.

Floor-system : Guard-rails, 6 in. × 8 in.

Ties, 7 in. × 8 in. × 10 ft.

Stringers, 6 in. × 14 in. × 30 ft. and 15 ft.

Corbels, 12 in. × 18 in. × 6 ft.

Bent: Cap, 12 in. × 12 in. × 12 ft.

Posts, 12 in. × 12 in.

Intermediate caps, 12 in. × 12 in.

Sway-braces, 2 in. × 12 in.

Sill, 12 in. × 12 in.

Longitudinal braces : Horizontal, 6 in. × 12 in. × 16 ft.

Diagonal, 6 in. × 8 in.

Sub-sills, 12 in. × 12 in. × 6 ft.

Dimensions of Iron Details.

Bolts : ¾ in. × 15 in. ; guard-rails to ties.

¾ in. × 19 in. ; stringers to corbels.

¾ in. × 21 in. ; stringer-joints ; packing-bolts.

¾ in. × 27 in. ; longitudinal braces to posts, and post-caps to intermediate caps.

—— × 15 in. ; intersection of diagonal longitudinal braces.

Drift-bolts : ¾ in. × ¾ in. × 18 in. ; corbels to cap.

Spikes : Boat, $\frac{7}{16}$ in. × 10 in. ; ties to stringers.

Cut 50-penny ; bracing to posts.

Cast washers : Under head and nut of each bolt.

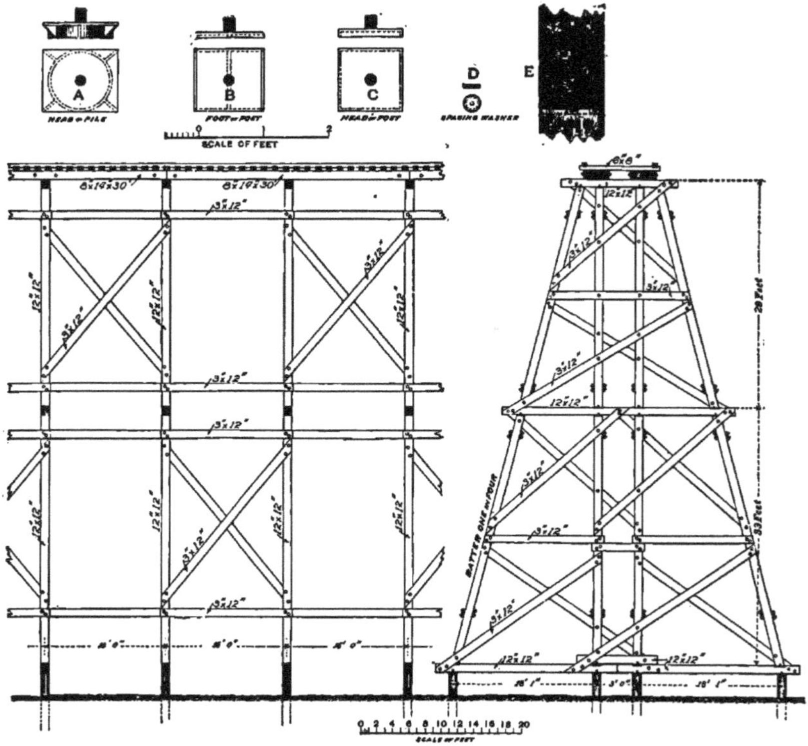

PLATE XXII.—HIGH FRAMED TRESTLES, MINNEAPOLIS & ST. LOUIS RAILWAY.

HIGH FRAMED TRESTLES, MINNEAPOLIS & ST. LOUIS RAILWAY.—PLATE XXII.

(See also Plate V.)

Dimensions of Timbers.

Floor-system : Guard-rails, 6 in. × 8 in.

Ties, 6 in. × 8 in. × 10 ft.

Stringers, 8 in. × 14 in. × 30 ft.

Bent : Cap, 12 in. × 12 in. × 14 ft.

Plumb-posts, 12 in. × 12 in.

Batter-posts, 12 in. × 12 in.

Sway-bracing : Horizontal, 3 in. × 12 in.

Diagonal, 3 in. × 12 in.

Intermediate cap, 12 in. × 12 in.

Sill, 12 in. × 12 in.

Sill-splice, 12 in. × 12 in.

Longitudinal braces : Horizontal, 3 in. × 12 in.

Diagonal, 3 in. × 12 in.

Dimensions of Iron Details.

Bolts : $\frac{3}{4}$ in. × ——; stringer-joints; packing-bolts.

—— × ——; braces to posts.

Lag-screws : —— × ——; stringer-brackets to caps.

Spikes, boat : $\frac{1}{2}$ in. × 8 in.; guard-rails to ties.

Drift-bolts : $\frac{5}{8}$ in. × 12 in. ; ties to stringers.

Cast separators : Between stringer-pieces.

Cast washers : Under head and nut of each bolt.

Cast brackets : Stringers to caps.

Cast pile-caps : ⎫

Cast post-caps : ⎬ As per details.

Cast post foot-blocks : ⎭

For complete description, etc., of this trestle see *Railroad Gazette*, April 17, 1891.

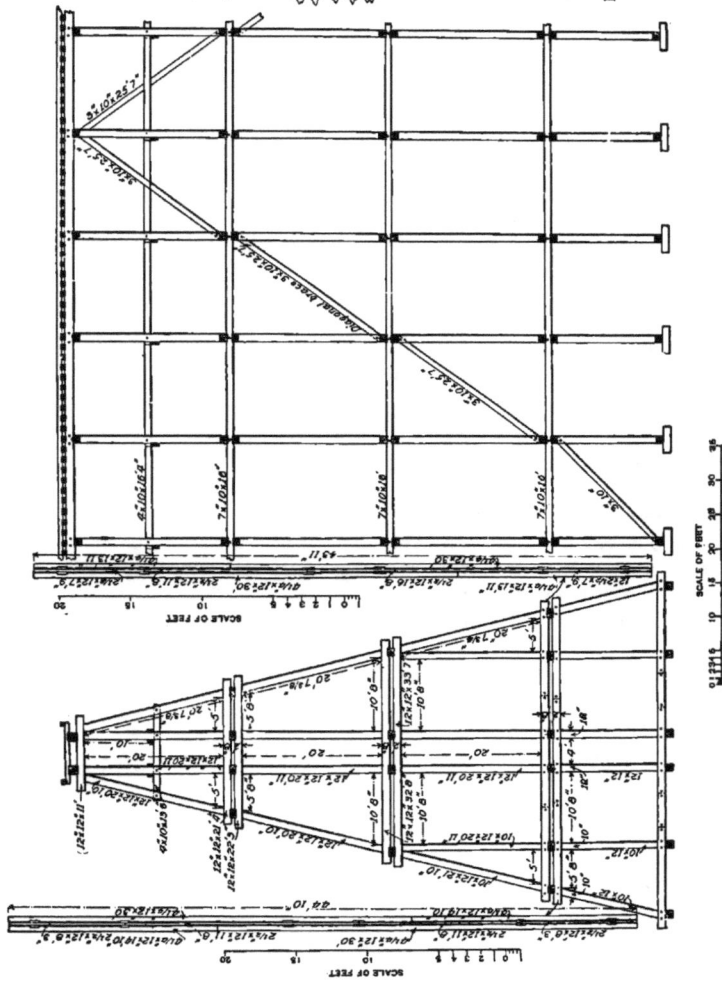

FIG. 3.—DETAILS OF SPLIT-CAP AND SILL-JOINTS.

FIG. 1.—CROSS-SECTION.

FIG. 2.—ELEVATION.

PLATE XXIII.—STANDARD FRAMED TRESTLE, GEORGIA PACIFIC RAILWAY.

124

STANDARD FRAMED TRESTLE, GEORGIA PACIFIC RAILWAY.—PLATE XXIII.

Dimensions of Timbers.

Floor-system : Guard-rails, 8 in. × 8 in.

                Ties, 8 in. × 10 in. × 9 ft.

                Stringers, 8 in. × 16 in.

Bent : Cap, 12 in. × 12 in. × 11 ft.

        Plumb-posts, 12 in. × 12 in.

        Batter-posts, 12 in. × 12 in., and 10 in. × 12 in.

        Vertical counter-posts, 10 in. × 12 in.

        Intermediate caps and sills, 12 in. × 12 in.

        Compound sills and caps, 4¼ in. × 12 in., and 2½ in. × 12 in.

Longitudinal bracing : Horizontal, 4 in. × 10 in. × 16 ft. 4 in., and 7 in. × 10 in. × 16 ft.

                    Diagonal, 3 in. × 10 in.

Sub sills, 12 in. × 12 in.

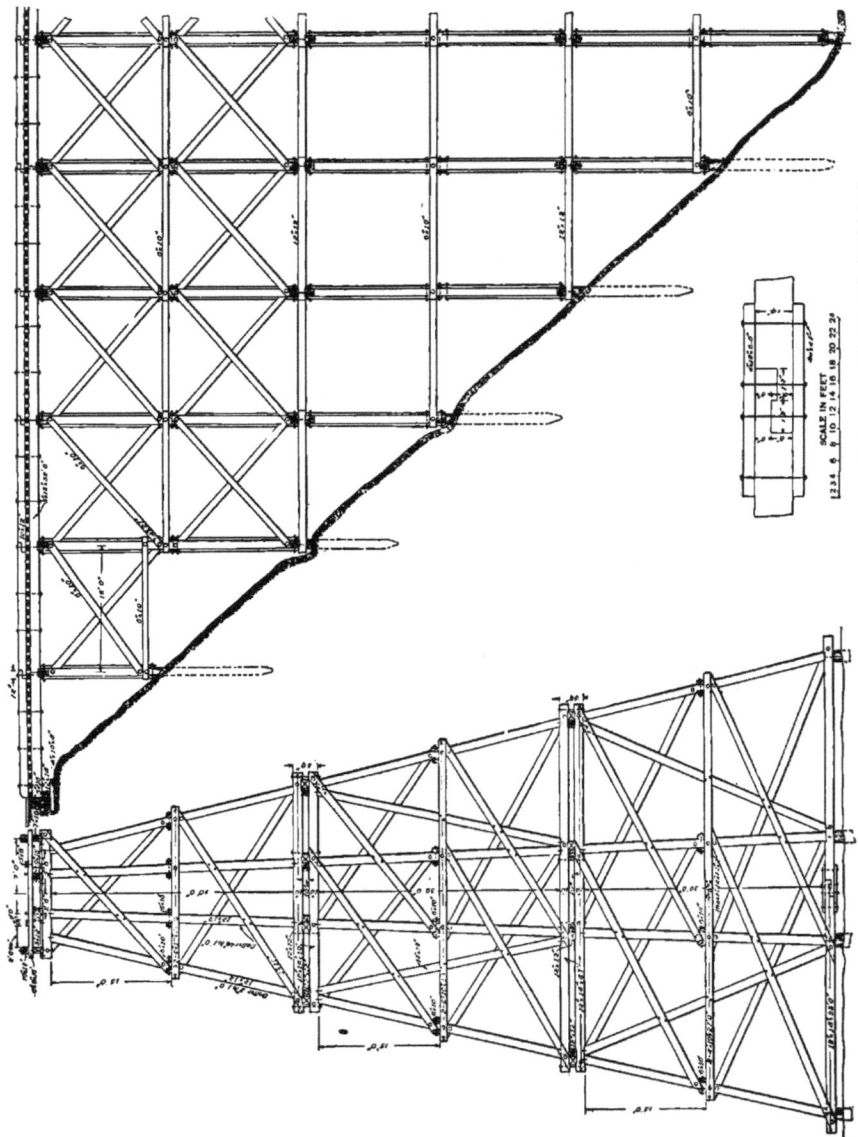

PLATE XXIV.—STANDARD FRAMED TRESTLE, OREGON & WASHINGTON TERRITORY RAILROAD

Standard Framed Trestles, Oregon & Washington Territory Railroad.—
Plate XXIV.

Dimensions of Timbers.

Floor-system: Guard-rails, 10 in. $\times$ 12 in., and 5 in. $\times$ 8 in.

Ties, 6 in. $\times$ 8 in. $\times$ 16 ft.

Track-stringers, 9 in. $\times$ 16 in. $\times$ 32 ft.

Jack-stringers, 7 in. $\times$ 16 in. $\times$ 32 ft.

Spreaders, 3 in. $\times$ 12 in.

Bent: Cap, 12 in. $\times$ 14 in. $\times$ 16 ft.

Plumb-posts, 12 in. $\times$ 12 in.

Batter-posts, 12 in. $\times$ 12 in.

Intermediate caps and sills, 12 in. $\times$ 14 in.

Sway-bracing: Horizontal, 4 in. $\times$ 10 in.

Diagonal, 4 in. $\times$ 10 in.

Main sill, 12 in. $\times$ 14 in.

Longitudinal bracing: Horizontal, 6 in. $\times$ 10 in.

Diagonal, 6 in. $\times$ 10 in.

Purlins, 12 in. $\times$ 12 in. $\times$ 18 ft.

Dimensions of Iron Details.

Bolts: $\frac{3}{4}$ in. $\times$ $50\frac{1}{2}$ in.; floor-system to caps.

$\frac{3}{4}$ in. $\times$ 41 in.; sills to caps of different decks.

$\frac{3}{4}$ in. $\times$ 37 in.; outside guard-rails to jack-stringers.

$\frac{3}{4}$ in. $\times$ 27 in.; $\Big\}$ longitudinal bracing.
$\frac{3}{4}$ in. $\times$ $24\frac{3}{4}$ in.;

$\frac{3}{4}$ in. $\times$ 23 in.; sway-brace splice, sill-splice, horizontal sway-bracing to posts.

$\frac{3}{4}$ in. $\times$ 22 in.; stringer-joints; packing-bolts.

$\frac{3}{4}$ in. $\times$ 19 in.; sway-braces to posts.

$\frac{5}{8}$ in. $\times$ 11 in.; inside guard-rails to ties.

Drift-bolts: $\frac{3}{4}$ in. $\times$ 24 in.; sill to piles.

Dowels: 1 in. $\times$ 6 in.; posts to caps and sills.

Spikes: Cut 60-penny; spreaders and brace-blocks to caps.

Boat, $\frac{1}{2}$ in. $\times$ 9 in.; sway-braces to posts.

Cast washers: Under head and nut of each bolt.

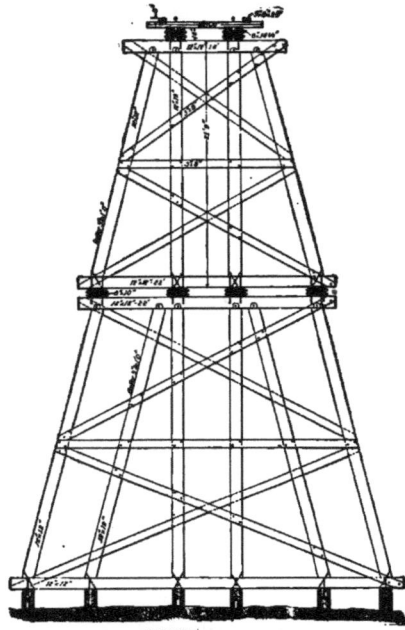

PLATE XXV.—STANDARD FRAMED TRESTLE, FORT WORTH & DENVER CITY RAILWAY

STANDARD FRAMED TRESTLE, FORT WORTH & DENVER CITY RAILWAY

PLATE XXV.

Dimensions of Timbers.

Floor-system: Guard-rails, 5 in. × 8 in. × 29 ft.
Ties, 6 in. × 8 in. × 10 ft.
Stringers, 8 in. × 14½ in.

Bent: Cap, 12 in. × 12 in. × 14 ft.
Plumb-posts, 12 in. × 12 in.
Batter-posts, 12 in. × 12 in.
Intermediate cap and sill, 12 in. × 12 in.

Sway-bracing: Horizontal, 3 in. × 8 in.
Diagonal, 3 in. × 8 in.

Main sill, 12 in. × 12 in.

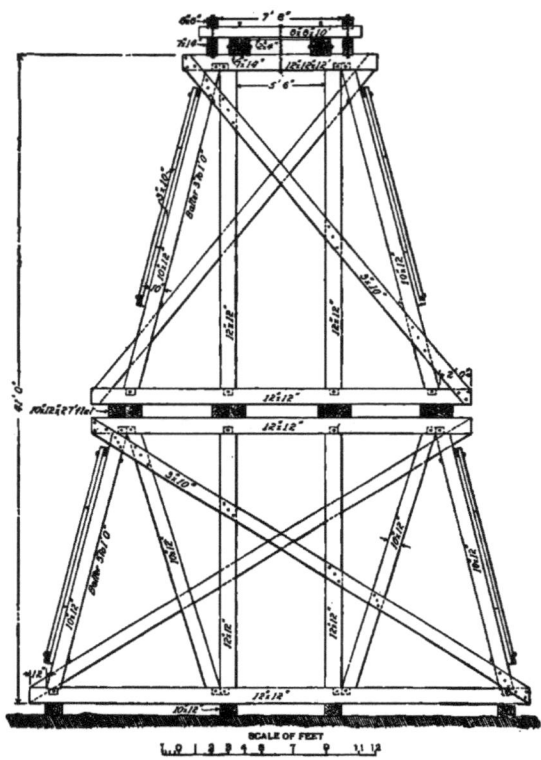

SCALE OF FEET

PLATE XXVI.—DOUBLE-DECK TRESTLE, RICHMOND & DANVILLE R. R.

130

STANDARD FRAMED TRESTLE, RICHMOND & DANVILLE RAILROAD.—PLATE XXVI.

Dimensions of Timbers.

Floor-system : Guard-rails, 8 in. × 8 in.
                    Ties, 8 in. × 8 in. × 10 ft.
                    Stringers, 7 in. × 14 in.
                    Spreader, 2 in. × 4 in. × 3 ft. 9 in.

Bent: Cap, 12 in. × 12 in. × 12 ft.
       Plumb-posts, 12 in. × 12 in.
       Batter-posts, 10 in. × 12 in.
       Counter-posts, 10 in. × 12 in.
       Intermediate sills and caps, 12 in. × 12 in.
       Sway-braces, 3 in. × 10 in.
       Main sill, 12 in. × 12 in.

Purlins, 10 in. × 12 in. × 27 ft.
Sub-sills, 10 in. × 12 in.
Longitudinal braces, 3 in. × 10 in.

Dimensions of Iron Details.

Bolts : —— × —— ; guard-rails to jack-stringers.
       $\frac{4}{8}$ in. × —— ; stringer-joints ; packing-bolts.
       —— × 36 in. ; floor-system to caps.
       $\frac{7}{8}$ in. × —— ; longitudinal bracing to posts.
Spikes : —— × 7 in. ; sway-braces to posts, etc.
        —— ; spreaders to ties.
Cast washers : Under head and nut of each bolt.

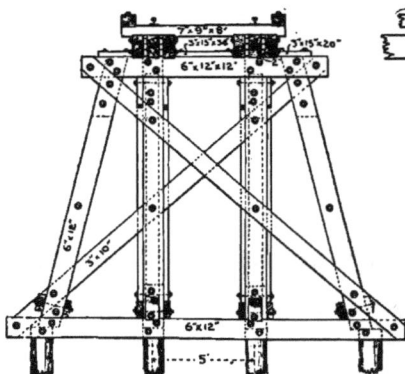

FIG. 1.—CROSS-SECTION.

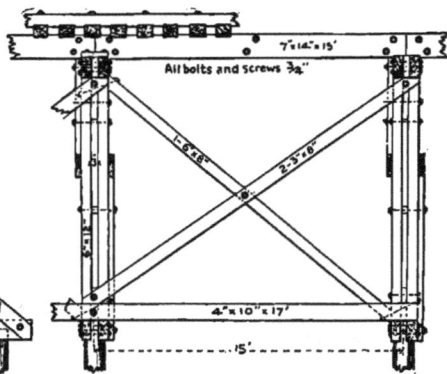

FIG. 2.—ELEVATION.

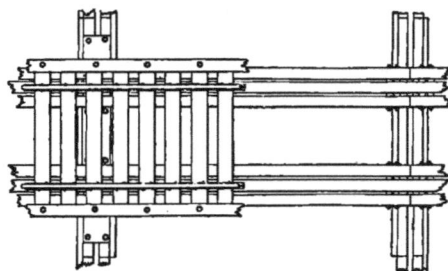

FIG. 3.—PLAN.

PLATE XXVII.—STANDARD FRAMED TRESTLE, CLEVELAND & CANTON RAILROAD.

132

STANDARD FRAMED TRESTLE, CLEVELAND & CANTON RAILROAD.—PLATE XXVII.

Dimensions of Timbers.

Floor-system : Guard-rails, 8 in. × 8 in., notched 1 in. over ties.

Ties, 7 in. × 9 in. × 8 ft., notched 1 in. over stringers.

Stringers, 7 in. × 14 in. × 15 ft., notched 1 in. over caps.

Brace-blocks, $\begin{cases} 3 \text{ in.} \times 15 \text{ in.} \times 20 \text{ in.} \\ 3 \text{ in.} \times 15 \text{ in.} \times 34 \text{ in.} \end{cases}$

Bents : Caps, 6 in. × 12 in. × 12 ft.

All posts, 6 in. × 12 in.

Sills, 6 in. × 12 in.

Sway-braces, 3 in. × 10 in.

Tenon-blocks, 3 in. × 12 in. × 3 ft.

Longitudinal braces : Girts, 4 in. × 10 in. × 17 ft.

Diagonals, $\begin{cases} 6 \text{ in.} \times 8 \text{ in.} \\ 3 \text{ in.} \times 8 \text{ in.} \end{cases}$

Dimensions of Iron Details.

Bolts : ¾ in. × 18 in.; post, sill, and cap ; packing-bolts.

¾ in. × 28 in.; stringer-joints ; packing-bolts.

¾ in. × 21 in.; sway-braces to posts.

¾ in. × ——; diagonal longitudinal braces to pos

¾ in. × 17 in.; diagonal longitudinal braces to posts ; intersection of diagonals.

Lag-screws : ¾ in. × ——, $\begin{cases} \text{guard-rails to ties.} \\ \text{brace-blocks to caps.} \end{cases}$

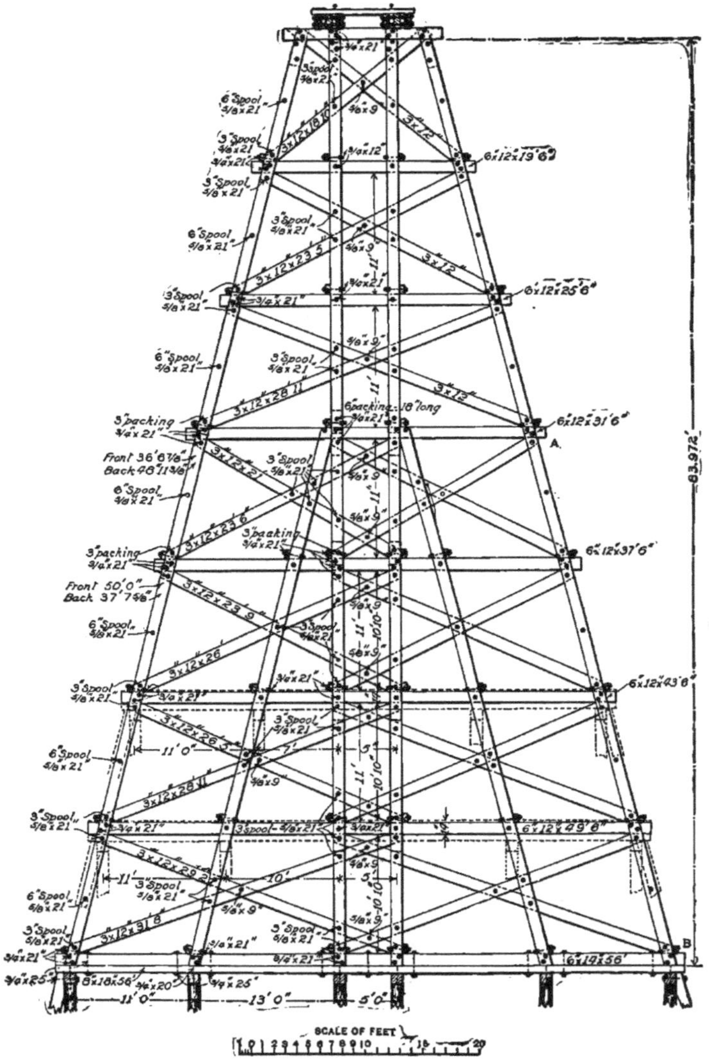

FIG. 1.—CROSS-SECTION.

PLATE XXVIII.—STANDARD TRESTLE, CALIFORNIA CENTRAL RAILWAY.

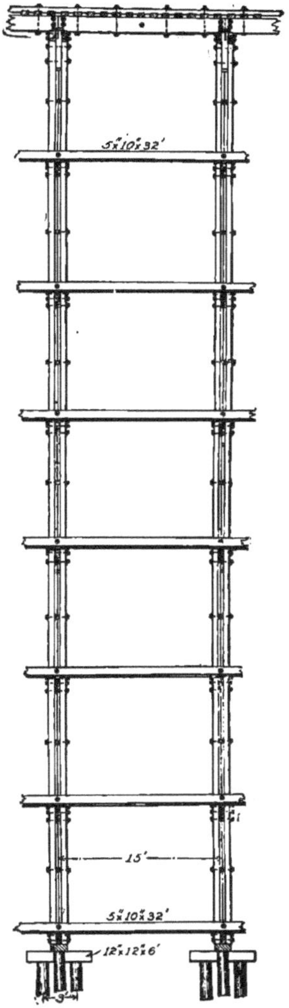

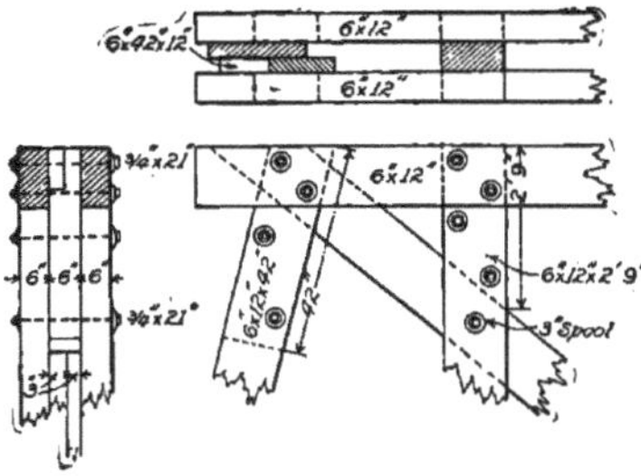

FIG. 3.—CAP AND POST JOINT.

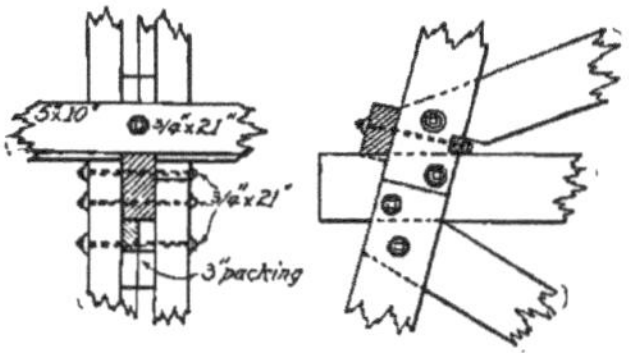

FIG. 4.—POST-SPLICE.

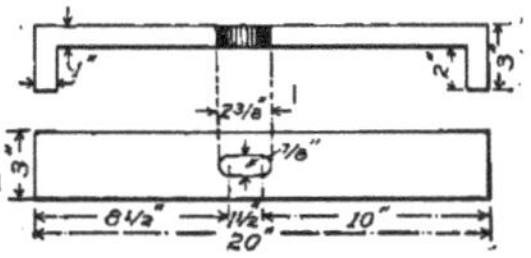

FIG. 5.—STRAP.

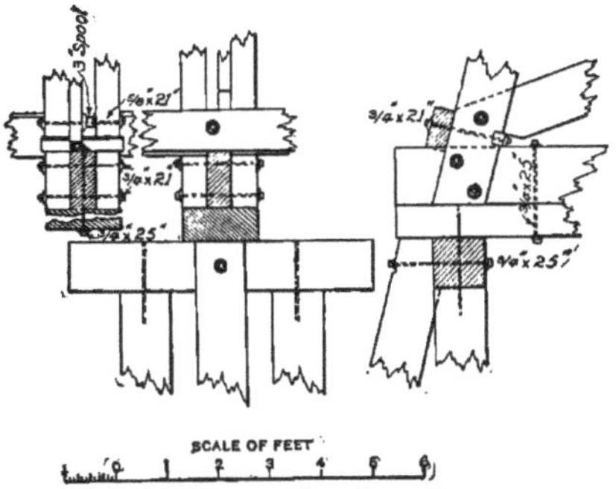

SCALE OF FEET

FIG. 6.—SILL AND POST JOINT DETAILS.

FIG. 2.—ELEVATION.

PLATE XXIX.—STANDARD TRESTLE, CALIFORNIA CENTRAL RAILWAY.

135

STANDARD FRAMED TRESTLE, CALIFORNIA CENTRAL RAILWAY. PLATES XXVIII AND XXIX.

Dimensions of Timbers.

Floor-system: Guard-rails, 6 in. × 8 in., notched.

Ties, 6 in. × 8 in. × 9 ft., notched.

Stringers, 8 in. × 16 in. × 30 ft.

Bents: Caps, 6 in. × 12 in. × 14 ft.

All posts, 6 in. × 12 in.

Intermediate caps, 6 in. × 12 in.

Diagonal sway-braces, 3 in. × 12 in.

Sill, 6 in. × 14 in.

Under part of sill, 8 in. × 18 in.

Pile-caps, 12 in. × 12 in. × 6 ft.

Packing-blocks between posts and cap-pieces, 6 in. × 12 in. × 42 in., and 6 in. × 12 in. × 2 ft. 9 in.

Longitudinal braces, 5 in. × 10 in. × 32 ft.

Dimensions of Iron Details.

Bolts: Guard-rails to stringers.

Stringer-joints; packing-bolts.

⅞ in. × 21 in.; sway-braces to posts.

⅞ in. × 9 in.; intersection of sway-braces.

¾ in. × 21 in.; intermediate caps to posts; girts to posts.

¾ in. × 25 in.; two parts of sill together; brace-pile to pile-caps.

Drift-bolts: ¾ in. × 20 in. sill to pile-caps; caps to piles.

Cast washers: Under head and nut of each bolt.

Cast separators or spools: 3 in. × —— and 6 in. × ——.

Cast strap: 20 in. × 3 in. × 1 in. as per detail; girts to posts.

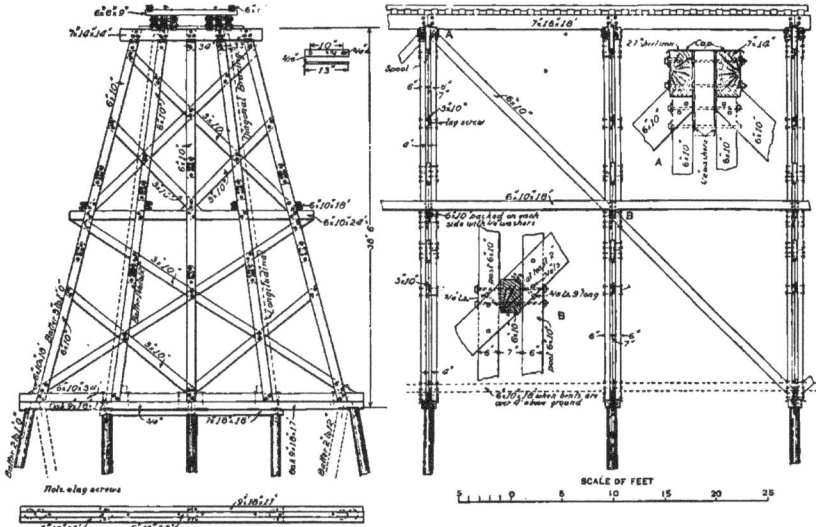

PLATE XXX.—STANDARD FRAMED TRESTLE, TOLEDO, ST. LOUIS & KANSAS CITY RAILROAD.

STANDARD FRAMED TRESTLE, TOLEDO, ST. LOUIS & KANSAS CITY RAILROAD.
PLATE XXX.

Dimensions of Timbers.

Floor-system : Guard-rails, 6 in. × 6 in. × 18 ft.
     Ties, 6 in. × 8 in. × 9 ft.
     Stringers, 7 in. × 18 in. × 18 ft.
     Spreader, 3 in. × 12 in. × 3 ft.
     Brace-blocks, 3 in. × 12 in. × 15 in.
Bent : Cap, 7 in. × 14 in. × 14 ft.
   Plumb-post, 6 in. × 10 in.
   Inclined posts, 6 in. × 10 in.
   Splice-blocks, 6 in. × 10 in. × 2 ft.
   Sway-bracing : Horizontal, 6 in. × 10 in.
        Diagonal, 3 in. × 10 in.
   Sill, 6 in. × 10 in., 9 in. × 18 in., and 7 in. × 18 in.
Longitudinal bracing : Horizontal, 6 in. × 10 in. × 18 ft.
       Diagonal, 6 in. × 10 in.

Dimensions of Iron Details.

Bolts : ⅝ in. × 31 in. ; stringer-joints : packing-bolts.
   ⅝ in. × 23 in. ; cap-pieces together.
   ⅝ in. × 21 in. ; post-splices, sway-brace intersections, posts to tenon-blocks, posts
     to sill.
Lag-screws : ¾ in. × 9 in. ; sway-braces to posts, longitudinal braces to posts, etc. ; spreader
     and brace blocks to cap.
     ¾ in. × 14 in. ; sill-pieces together.
Drift-bolts : ¾ in. × ¾ in. × 20 in. ; sill to piles.
Cast-separators : 4 in. thick } for ⅝-in. bolts.
      ½ in. " }
Cast-washers : Under head and nut of each bolt.
Splice-plates : ₁₆⁵ in. × 2 in. × 13 in.; stringer-joints.
Sheet-iron : No. 27, 30 in. × 36 ft. ; covering stringers.
    No 27, 24 in. × 14 ft. ; covering caps.
    No. 27, —— × 34 ft. ; covering sill.
    Also, sheet-iron to cover all places where fire can lodge.

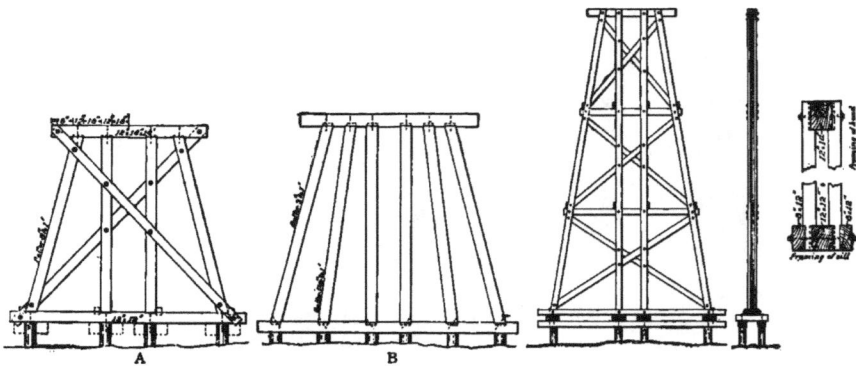

FIG. 1.—FRAMED BENTS.   B   FIG. 2.—40 TO 90 FT. HIGH.

PLATE XXXI.—STANDARD FRAMED TRESTLES, ATLANTIC & PACIFIC RAILROAD.

STANDARD FRAMED TRESTLE, ATLANTIC & PACIFIC RAILRCAD.—PLATE XXXI.
(See also Plate III.)

Dimensions of Timbers.

Floor-system : See Plate III.

Bent—Fig. 1 : Cap, 12 in. × 14 in. × 14 ft.
Plumb-posts, 12 in. × 12 in.
Batter-posts, 12 in. × 12 in.
Sway-braces, 3 in. × 10 in.
Sill, 12 in. × 12 in.

Fig. 2 : Cap, 12 in. × 14 in.
Plumb-posts, 6 in. × 12 in.
Batter-posts, 6 in. × 12 in.
Intermediate caps, 6 in. × 12 in.
Sway-braces, 3 in. × 10 in.
Sill, 8 in. × 12 in., and 12 in. × 12 in.

Sub-sills, 12 in. × 12 in.

Pile-caps, 12 in. × 14 in.

Dimensions of Iron Details.

See Plate III.

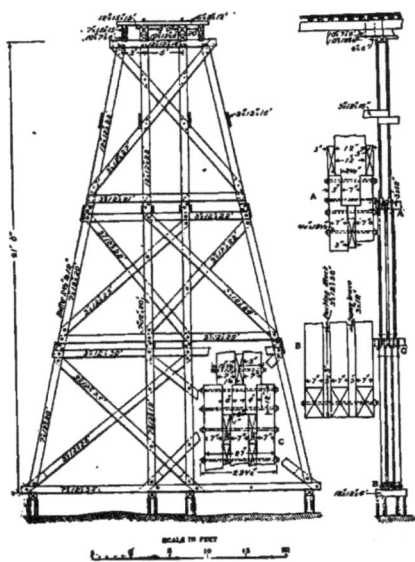

FIG. 1.—DOUBLE AND TRIPLE DECK FRAMED TRESTLE.

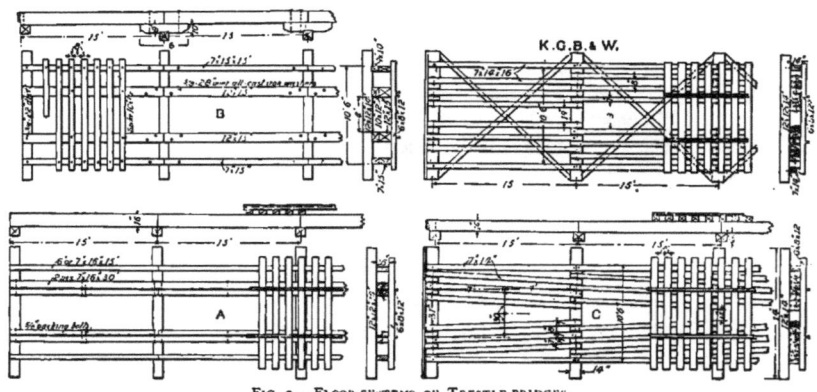

FIG. 2.—FLOOR-SYSTEMS OF TRESTLE-BRIDGES.

PLATE XXXII.—STANDARD FRAMED TRESTLE, MILWAUKEE & NORTHERN RAILROAD.

STANDARD FRAMED TRESTLE, MILWAUKEE & NORTHERN RAILROAD.—PLATE XXXII.

Dimensions of Timbers.

Floor-systems—A : Ties, 6 in. × 8 in. × 12 ft.
  Track-stringers, 7 in. × 16 in. × 30 ft.
  Jack-stringers, 6 in. or 7 in. × 16 in. × 15 ft.
  B : Ties, 6 in. × 8 in. × 12 ft.
  Track-stringers, 12 in. × 15 in. × 15 ft.
  Jack stringers, 7 in. × 15 in. × 15 ft.
  Track-stringer corbels, 10 in. × 12 in. × 6 ft.
  Jack-stringer corbels, 7 in. × 10 in. × 6 ft.
  C : Ties, 6 in. × 8 in. × 12 ft.
  Track-stringers, 7 in. × 14 in. × 16 ft.
  Jack-stringers, 7 in. × 14 in. × 15 ft.
  K., G. B. & W. R. R. :
  Ties, 6 in. × 8 in. × 12 ft.
  Track-stringers, 7 in. × 14 in. × 16 ft.
  Jack-stringers, 7 in. × 14 in. × 16 ft.
Bent : Cap, 12 in. × 12 in. × 14 ft.
  Plumb-posts, 12 in. × 12 in., and 7 in. × 12 in.
  Batter-posts, 12 in. × 12 in., and 7 in. × 12 in.
  Intermediate caps and sills, 3 in. × 12 in.
  Sway-braces, 3 in. × 12 in.
  Main sill, 7 in. × 12 in.
  Packing-blocks, 3 in. × 12 in. × 3 ft.
Pile-caps : 12 in. × 12 in. × 4 ft.
Longitudinal braces : 3 in. × 12 in. × 16 ft.

Dimensions of Iron Details.

Bolts : $\frac{3}{4}$ in. × 19$\frac{1}{2}$ in.
  $\frac{3}{4}$ in. × 22$\frac{1}{4}$ in.
  $\frac{3}{4}$ in. × 29$\frac{1}{4}$ in.
Spikes : —— × ——.

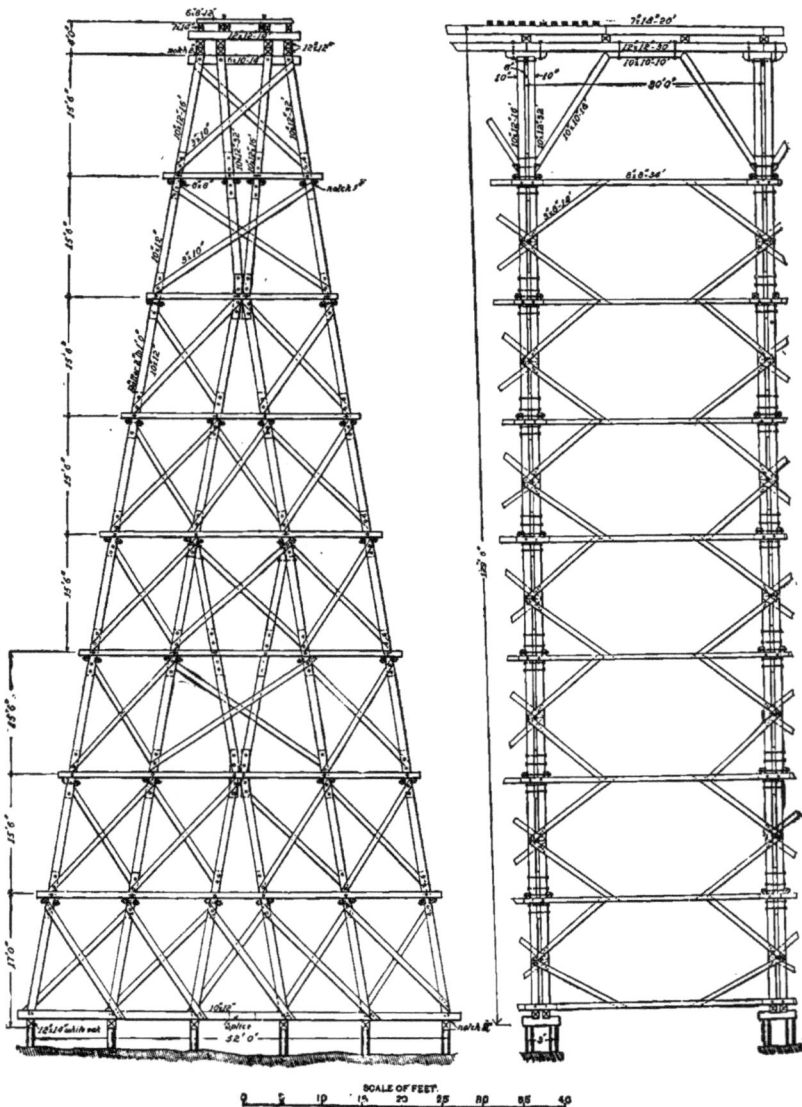

FIG. 1.—CROSS-SECTION.   FIG. 2.—ELEVATION.

PLATE XXXIII.—FRAMED TRESTLE, ST. PAUL, MINNEAPOLIS & MANITOBA RAILROAD.

FRAMED TRESTLES, ST. PAUL, MINNEAPOLIS & MANITOBA RAILROAD.—
PLATES XXXIII AND XXXIV.

Dimensions of Timbers.

FIGS. 1 AND 2.

Floor-system : Ties, 6 in. × 8 in. × 12 ft.

Track-stringers, 7 in. × 14 in. × 20 ft.

Jack-stringers, 7 in. × 14 in. × 20 ft.

Floor-beams, 12 in. × 12 in. × 14 ft.

Sub-stringers, 12 in. × 12 in. × 30 ft.

Corbels, 12 in. × 12 in. × 5 ft.

Bent : Cap, 6 in. × 10 in. × 14 ft.

Posts, 10 in. × 12 in.

Sway-bracing : Horizontal, 8 in. × 8 in.

Diagonal, 3 in. × 10 in.

Splice-blocks, 6 in. × 12 in. × 6 ft.

Sill, 10 in. × 12 in.

Pile-caps, 12 in. × 14 in.

Knee-braces : Top cord, 10 in. × 10 in. × 10 ft.

Diagonals, 10 in. × 10 in. × 18 ft.

Longitudinal braces : Horizontal, 8 in. × 8 in. × 34 ft.

Brackets, 3 in. × 8 in. × 14 ft.

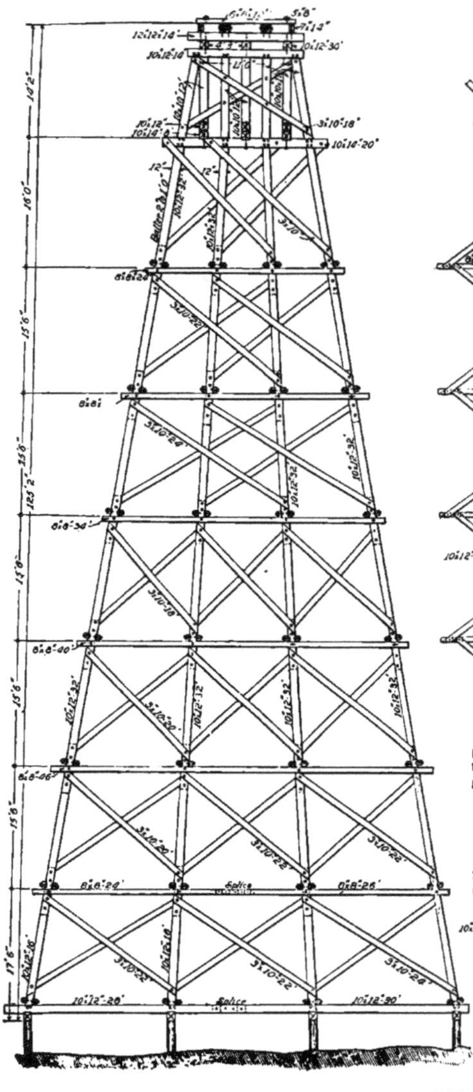

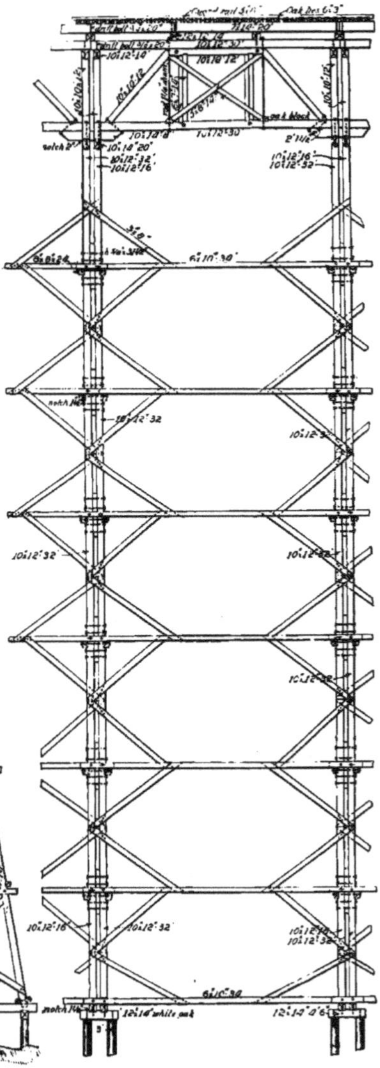

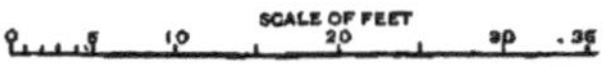

FIG. 3.—CROSS-SECTION.   FIG. 4.—ELEVATION.

PLATE XXXIV.—FRAMED TRESTLE, ST. PAUL, MINNEAPOLIS & MANITOBA RAILROAD.

FIGS. 3 AND 4.

Floor-system : Guard-rails, 5 in. × 8 in.

Ties, 6 in. × 8 in. × 12 ft.

Track-stringers, 7 in. × 14 in. × 20 ft.

Jack-stringers, 7 in. × 14 in. × 20 ft.

Floor-beams, 12 in. × 12 in. × 14 ft.

Sub-stringers, 10 in. × 12 in. × 30 ft.

Bent : Upper cap, 10 in. × 12 in. × 14 ft.

Lower cap, 10 in. × 14 in. × 20 ft.

Upper posts, 10 in. × 10 in.

Main posts, 10 in. × 12 in.

Sway-bracing : Horizontal, 8 in. × 8 in.

Diagonal, 3 in. × 10 in.

Splice-blocks, 6 in. × 12 in. × 6 ft.

Sill, 10 in. × 12 in.

Sway-brace splice-block, 6 in. × 8 in. × 5 ft.

Sill splice-block, 6 in. × 12 in. × 4 ft.

Pile-caps, 12 in. × 14 in. × 4 ft. 6 in.

Longitudinal bracing : Horizontal, 6 in. × 10 in. × 34 ft.

Brackets, 3 in. × 8 in.

Floor trusses : Upper chord, 10 in. × 10 in. × 12 ft.

Lower chord, 10 in. × 12 in. × 30 ft.

End-posts, 10 in. × 10 in. × 12 ft.

Diagonals, 5 in. × 8 in. × 14 ft.

Lateral braces, 6 in. × 7 in. × 14 ft.

Foot-blocks : Corbels, 10 in. × 14 in. × 8 ft.

Dimensions of Iron Details.

Bolts : $\frac{5}{8}$ in. × 12$\frac{1}{2}$ in. ; guard-rails to ties.

$\frac{7}{8}$ in. × 17$\frac{1}{2}$ in. ; stringer-joints ; packing-bolts.

$\frac{3}{4}$ in. × 27$\frac{1}{2}$ in. ; longitudinal braces to posts.

$\frac{7}{8}$ in. × 28$\frac{1}{4}$ in. ; post-joints.

$\frac{5}{8}$ in. × 31$\frac{1}{2}$ in. ; diagonal sway-braces to posts.

$\frac{5}{8}$ in. × 41$\frac{1}{2}$ in. ; horizontal sway-braces to posts.

Drift-bolts : $\frac{3}{4}$ in. × 20 in. ; stringers to floor-beams ; floor-beams to sub-stringers ; sub-stringers to caps ; main sill to pile-caps ; pile-caps to piles.

Spikes : Boat, $\frac{1}{2}$ in. × 10 in. ; ties to stringers.

$\frac{5}{8}$ in. × 7 in. ; sway-bracing to posts ; bracket-braces to posts ; and longitudinal bracing.

Iron in trusses : Rods, 1$\frac{1}{4}$ in. × 11 ft. 4 in. ; between upper and lower chords.

Tie-rods, —— × —— ; three trusses together.

Bolts : —— × —— ; intersection of panel diagonals.

—— × 2 ft. 11 in. ; end-posts to lower chords.

—— × 3 ft. 5$\frac{1}{2}$ in. ; lower chords and corbels to caps.

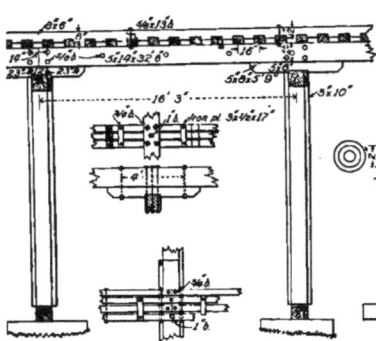

FIG. 3.—STRINGER JOINT.

FIG. 1.—ELEVATION.

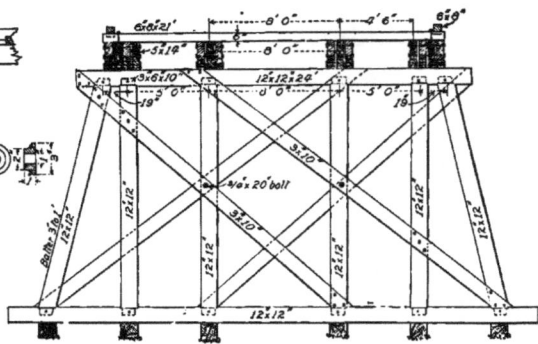

SCALE OF FEET

FIG. 2.—CROSS-SECTION.

PLATE XXXV.—DOUBLE-TRACK FRAMED TRESTLE, NEW YORK, WOODHAVEN & ROCKAWAY RAILROAD.

146

DOUBLE-TRACK FRAMED TRESTLE, NEW YORK, WOODHAVEN & ROCKAWAY RAILROAD.—
PLATE XXXV.

Dimensions of Timbers.

Floor-system : Guard-rails, 8 in. × 6 in.
　　　　　　Ties, 6 in. × 8 in. × 21 ft.
　　　　　　Stringers, 5 in. × 14 in. × 32 ft. 6 in.
　　　　　　Corbels, 5 in. × 8 in. × 5 ft. 9 in.
Bent : Cap, 12 in. × 12 in. × 24 ft.
　　　　Plumb-posts, 12 in. × 12 in.
　　　　Batter-posts, 12 in. × 12 in.
　　　　Sway-braces, 3 in. × 10 in.
　　　　Sill, 12 in. × 12 in.
　　　　Sub-sills,

Dimensions of Iron Details.

Bolts : $\frac{5}{8}$ in. × 13 in. ; guard-rails to ties.
　　　$\frac{4}{5}$ in. × —— in. ; stringer-joints ; packing-bolts.
　　　$\frac{3}{4}$ in. × —— in. ; stringers to corbels.
　　　$\frac{3}{4}$ in. × —— ; $\left.\right\}$ stringers to caps.
　　　1 in. × —— ; $\left.\right\}$
Plates : $\frac{1}{2}$ in. × 3 in. × 17 in. ; corbel-bolts.
Spikes : Ties to stringers.
Cast-washers : 1 in. × 3 in. ; under head and nut of each bolt.

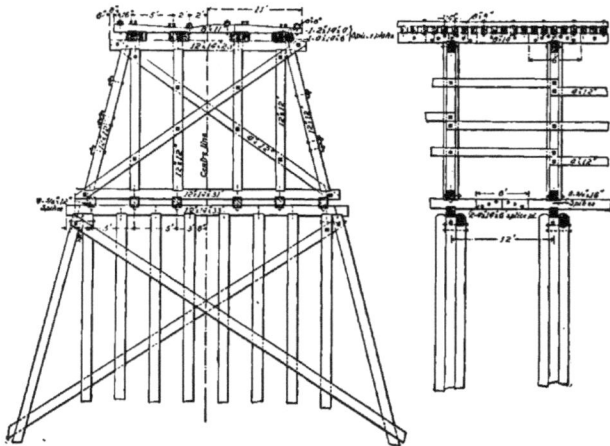

FIG. 1.—SECTIONAL AND SIDE ELEVATIONS, WINTHROP'S COVE TRESTLE.

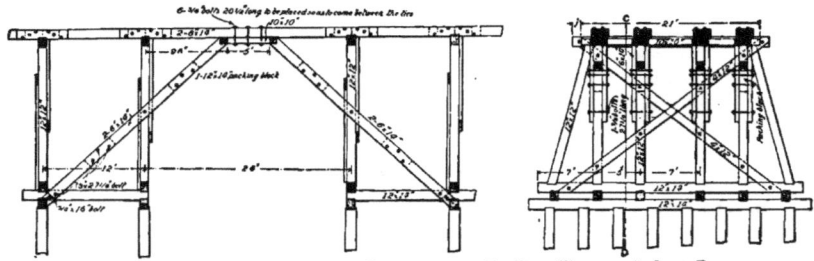

FIG. 2.—LONGITUDINAL AND TRANSVERSE SECTIONS OF 24 FT. BAY, WINTHROP'S COVE TRESTLE.

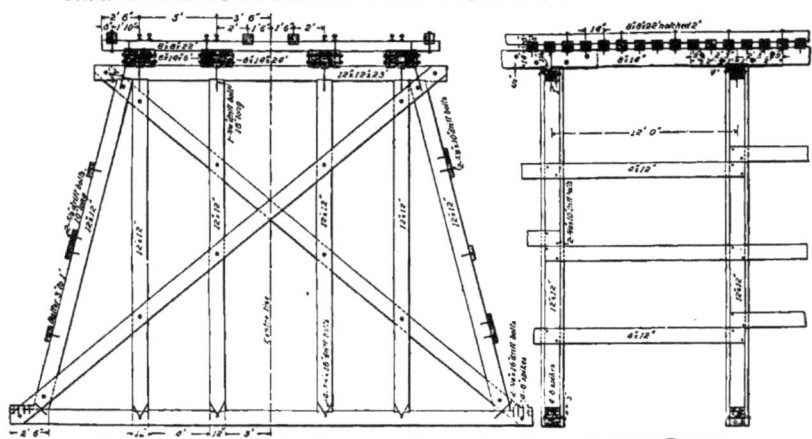

FIG. 3.—SECTIONAL AND SIDE ELEVATIONS, THAMES RIVER BRIDGE APPROACH TRESTLE.

PLATE XXXVI.—TRESTLE PLANS, NEW YORK, PROVIDENCE & BOSTON RAILROAD.

FRAMED TRESTLES, NEW YORK, PROVIDENCE & BOSTON RAILROAD.—
PLATE XXXVI.

FIGS. 1 AND 2 : WINTHROP'S COVE TRESTLE, ON 8° 15′ CURVE AND 0.714 GRADE.
FIG. 3 : THAMES RIVER BRIDGE APPROACH.

Dimensions of Timbers.

Floor-system : Guard-rails, 8 in. × 8 in.

Ties, 8 in. × 11 in. × 22 ft.

Stringers, 8 in. × 14 in. × 24 ft.

Splice-blocks, $\begin{cases} 2 \text{ in. } \times 14 \text{ in. } \times 4 \text{ ft.} \\ 4 \text{ in. } \times 14 \text{ in. } \times 6 \text{ ft.} \end{cases}$

Bent : Cap, 12 in. × 14 in. × 23 ft.

Plumb-posts, 12 in. × 12 in.

Batter-posts, 12 in. × 12 in.

Sway-braces, 4 in. × 12 in.

Sill, 12 in. × 14 in.

Longitudinal braces, 4 in. × 12 in. × 25 ft.

Purlins, 12 in. × 14 in.

Purlin splice-blocks, 4 in. × 14 in. × 6 ft.

Foundation : Pile-cap, 12 in. × 14 in.

Piles, 12 in. diameter.

Brace-piles, 12 in. diameter.

Knee-braces : Upper chord, 10 in. × 10 in. × 5 ft.

Straining-beams, 10 in. × 10 in. × 21 ft.

Diagonals, 6 in. × 14 in.

Splice-block, 12 in. × 14 in. × 4 ft.

Dimensions of Iron Details.

Bolts : $\frac{7}{8}$ in. × 15$\frac{1}{2}$ in. ; $\big)$
$\frac{7}{8}$ in. × 19$\frac{1}{2}$ in. ; $\big\}$ guard-rails to ties.

$\frac{3}{4}$ in. × 18$\frac{1}{4}$ in. sway-braces to posts, etc.

$\frac{7}{8}$ in. × 22 in. ; purlin splice.

$\frac{7}{8}$ in. × 26$\frac{1}{2}$ in. ; upper chord to stringers.

$\frac{3}{4}$ in. × 16 in. ; foot of knee-brace to prevent splitting.

$\frac{7}{8}$ in. × 27$\frac{1}{4}$ in. ; knee-braces to purlins, to packing-blocks, to posts.

1 in. × 18 in. ; longitudinal braces to posts.

1 in. × 22$\frac{1}{2}$ in. ; $\big\}$ stringer-joints ; packing-bolts.
1 in. × 26$\frac{1}{2}$ in. ; $\big\}$

1$\frac{1}{8}$ in. × 34 in. ; batter and sway-brace piles to piles.

Drift-bolts : $\frac{3}{4}$ in. × $\frac{3}{4}$ in. × 21 in. ; stringers to caps.

1 in. × 18 in. ; sills to posts.

1 in. × 20 in. ; cap to posts ; pile-cap to piles.

Spikes : —— × —— ; ties to stringers.

$\frac{5}{8}$ in. × 12 in. ; purlins to pile-caps.

$\frac{3}{4}$ in. × 16 in. ; sills to purlins.

Washers, wrought : $\frac{3}{4}$-in. bolts ; guard-rail bolts.

4 in. × 4 in. × $\frac{5}{8}$ in. ; sway-brace bolts.

4 in. × 4 in. × $\frac{1}{2}$ in. ; longitudinal brace-bolts.

4 in. × 4 in. × $\frac{5}{8}$ in. ; sway-brace, pile, etc., bolts.

3 in. × $\frac{5}{8}$ in. ; purlin splice-bolts.

3$\frac{1}{2}$ in. × $\frac{1}{2}$ in. ; stringer-bolts.

Cast separators : 3$\frac{1}{2}$ in. × 1 in. ; between splice-blocks and stringers.

3$\frac{1}{2}$ in. × 4 in. ; between stringers, where there are no splice-blocks.

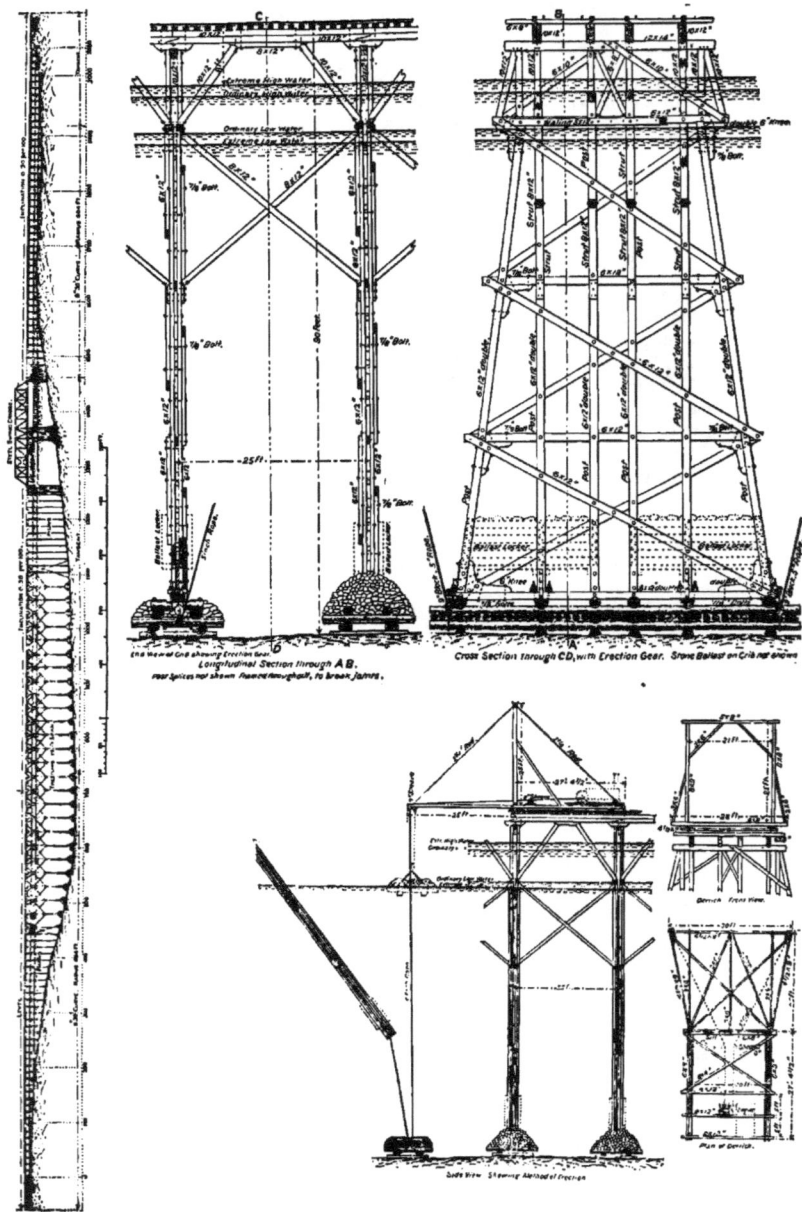

PLATE XXXVII.—DEEP-WATER FRAMED TRESTLE, INTERCOLONIAL RAILWAY.

DEEP-WATER FRAMED TRESTLE, INTERCOLONIAL RAILWAY.—PLATE XXXVII.

As the structure illustrated in Plate XXXVII is exceptional, and had to fulfil unusual requirements, it was thought best to reprint the full description of the work, as given in the *Railroad Gazette* of April 9, 1886.

It was designed to carry a short branch-line of minor importance across a narrow strait (the Narrows) in Halifax Harbor. The water being from 65 to 80 feet deep, some peculiar features of design and methods of construction were naturally required.

The branch as constructed (the Dartmouth Branch) is about 5 miles long. To avoid the trestle it would have been necessary to begin the branch at a point 9 miles or at one 14 miles distant, which would have made it seven or 12 miles long, and required a special train service in operating it.

Richmond yard being on the shore of the narrow passage between Halifax Harbor and Bedford Basin, at the most favorable point for bridging it, the structure shown was built instead, permitting the branch to leave the yard inside the semaphore, thus enabling the shunting-engines to do the business on it without in any way interfering with the traffic of the main line.

The Narrows are about 1500 ft. wide, and from 65 to 80 ft. deep in the channel where the line crosses for a distance of 650 ft. The mean rise and fall of the tide is 6 ft., causing a current through the Narrows of about 1½ miles an hour. At spring-tide, with a strong wind, this is sometimes increased to three miles an hour.

The bottom is generally compact gravel, mixed with stones and bowlders. In no place could a bar be driven more than 3 ft.; below that depth was apparently ledge-rock.

In severe winters ice forms in Bedford Basin, but owing to the extreme narrowness of the outlet into the harbor, it is held in the basin until decayed by the spring weather. The bridge has a total length of 2050 ft., of which 1204 ft. is on piling, 650 ft. trestling in the channel, and the remaining 196 ft. is a steel swing-bridge.

The piling, where in deep water, was well stiffened transversely by brace-piles, which were driven plumb and afterwards drawn over to a considerable angle, when they were fitted to the capping and bolted. The pivot pier for the swing-bridge is of masonry, and has a passage for vessels on each side of 85 ft. in the clear.

From the top of the pier to 2 ft. below low-water it is laid in cement, and is circular in form, with a diameter of 20 ft. Thence to the bottom, about 33 ft., it is built square, with a batter of 1 in 12, and is laid without mortar. Large stones only were permitted to be used in the square portion of the work, and were required to be full bedded throughout and closely fitted. Each course was carefully dressed and put together in the quarry upon a level platform; the stones were then marked with white paint at all connections with their fellows, and carefully numbered.

The courses were then forwarded to the site of the pier, where they were lowered from a lighter, each stone in its proper order, and received by a diver, who, standing on the course last laid, placed them in position, using lines, straight-edge, and spirit-level to insure all possible accuracy. A complete course was frequently laid in a day by the one diver employed, for with the footings once properly levelled he had but little to do to keep the work in good order. Before putting in the foundation courses the sloping bottom was properly benched by the diver, and frequent testings as the work proceeded showed that perfect line and level was being kept.

No accident or difficulty of any kind occurred in the construction of the pier, the work being carried on as smoothly and regularly as if in the open air; the steam winch of the lighter working with quickness and precision as the diver signalled his directions.

The time occupied in building the pier was 70 days, the same diver being employed throughout. The cost per cubic yard was $23.

The trestling across the channel consisted of timber bents, framed as shown on the accompanying drawings. The bents were placed 25 ft. apart between centres, and rested on a ballasted timber crib, which had previously been lowered in place. The bents were floated to the site and drawn down to their seat on the cribs by the methods shown in the cuts and described hereafter. The work of putting down the trestling was commenced on the west side of the channel August 8, 1884, and on the east side October 4, 1884. In all 25 bents were put down, in depths generally from 70 to 80 ft. The two sides were connected November 20,

1884. When the level portion of the channel was reached three bents were sometimes put down in a week. One diver, with occasionally an assistant, worked on each side of the channel. In addition to the travelling derrick shown, a lighter was provided for each side, having a steam-winch for lowering ballast, etc., and a steam-pump for the diver.

The correct centring at each bent was given by a theodolite placed at the outer end of the piling, and at slack-water lining in the rope holding the hammer of the floating pile-driver, which had been brought approximately into position, with the hammer raised about one foot from the bottom. When correctly lined the diver was signalled, and a bolt driven into the ground at the centre of the hammer.

It is not anticipated that there will be any trouble from worms, as the strength of the current and the large amount of fresh water discharged into the basin render their presence in the Narrows improbable. The wharves in the harbor also show that the nearer the Narrows are approached the less destructive are the worms. It is therefore hoped that the bents below low-water will but rarely require to be renewed, and they have been constructed of sawn hemlock, a cheap and sufficiently good material where secure from decay. The upper or supplementary bents were constructed of white pine, as more durable, and are so connected with the lower bents that, though erected as a whole, they can be easily separated and renewed.

The work of preparing the bottom for the crib foundation of the bent was as follows : Six flattened timbers 10 ft. long and weighted were lowered to the bottom. These were bedded by the diver, and were brought to a uniform level by means of a long straight-edge with spirit-level attached. Where the slope or character of the ground demanded, additional timbers were placed under these bed logs to bring them to the required height, the whole being filled in and about with stone. In fairly level ground the six logs could be bedded by one diver in 1½ days. In the worst cases, where the slope of the bottom was 1 in 2½ longitudinally and 1 in 5 transversely, it took the same diver six days to bed them properly.

The crib for each boat was next launched from the ways on which it was constructed, and floated out, and the lines from the winch on the travelling derrick attached to the chain at each end, by hooking on the iron swivel-blocks as shown.

The crib was supported until about nine tons of ballast had been thrown on, when it was lowered to its place on the bed-logs. When near the bottom the diver signalled any slight alteration required in its position, and the correction was made by side lines. The time occupied in lowering the cribs and finally adjusting them was about 1½ hours.

In difficult bottom the diver then proceeded to the next foundation, leaving an assistant to place the remainder of the ballast on the crib. This took about 1½ days to do properly.

The bent, which, like the crib, was built on ways on the shore, was next launched and towed to the site, and the lines from the travelling derrick, which passed through the blocks at the ends of the crib, were attached to the sill of the bent. About 10 tons of ballast were next placed in the lockers near the bottom, and the engine was started, drawing the bent gradually downward till, led by the blocks, it rested in its proper place on the crib. It was readily adjusted vertical by a line to the cap, and was then secured by bolting on temporary stays from the end of the bridge. The diver then permanently secured the bent to the crib in the manner shown, by lifting the galvanized-iron fastenings into place, fitting on the cover-blocks and screwing home the nuts. The fastenings were so arranged that they could be thrown back out of the way until the bent was finally settled in place.

The time occupied in towing out, hauling down, and adjusting a bent, together with the complete fitting and securing of the fastenings, was in general about 1½ days; of this the actual hauling down occupied but a small portion.

The permanent stringers were next placed and sleepered, the rails for the derricks laid, and the derrick run out for the next crib.

The average cost of the trestling per bent completed ready for the rails was as follows :

11 M. ft. B. M., hemlock,	@	$6.47	$71.17
9 M. ft. B. M., pine,	@	16.00	144.00
12 knees,	@	3.50	42.00
500 lbs. ordinary iron,	@	.04½	22.50
1800 lbs. galvanized iron,	@	.08	144.00
Crib—material and work,	@	60.00	60.00
Framing, 20 M. ft. B. M.,	@	10 00	200.00
Stone ballast, 66 tons,	@	.40	26.40
Diving work,	@	108.00	108.00
Incidentals,			31.93
Total,			$850.00

The bents number 25, making the total cost of the trestling $21,250, or at the rate of nearly $33 per lineal foot.

No accident of any kind occurred in putting down the trestle-bents or foundations, everything working smoothly throughout. All iron to be exposed to the action of salt water was galvanized. The crib foundations, from their position, and from being covered with stone, may be considered secure from the action of worms or other destroying agencies. Should a bent at any time require to be removed, it can be easily released from the crib and a new one substituted. In the deepest water the divers worked skilfully and without difficulty, and by coming to the surface for a few minutes every 1½ hours, were enabled to do good work throughout the entire day. All levelling and lining under water was accurately done, as proved when the bents were drawn down to their place. The divers were paid $150 per month each, with board; the assistant-divers about half that amount.

Steam-pumps were used for supplying air, in preference to those worked by hand, the increased regularity of stroke being of importance in deep water.

The current at the bottom, while not so rapid as at the surface, was more changeable, sometimes almost entirely ceasing and then suddenly recommencing. as though restrained temporarily by eddies or cross-currents. The divers, however, were rarely prevented by the current from working satisfactorily. Very severe gales occurred during the construction of the bridge and after its completion. No movement or working was at all perceptible during their continuance.

The bridge has now been completed and in operation nearly a year. Trains preceded by two locomotives crossing at 15 miles an hour have failed to produce the slightest motion or settlement in any part of the structure.

The work was planned and carried through under the direction of Mr. P. S. Archibald, Chief Engineer of the Intercolonial Railway.

PLATE XXXVIII.—STANDARD FRAMED TRESTLE, ESQUIMALT & NANAIMO RAILWAY.

STANDARD FRAMED TRESTLE, ESQUIMALT AND NANAIMO RAILWAY.—PLATE XXXVIII.

Dimensions of Timbers.

Floor-system : Guard-rails, 6 in. × 9 in.

      Ties, 8 in. × 9 in. × 13 ft.

      Stringers, 9 in. × 16 in.

Bents : Caps, 12 in. × 12 in. × 16 ft.

      Plumb-posts, 12 in. × 12 in., and 12 in. × 14 in.

      Batter-posts, 12 in. × 12 in., and 12 in. × 14 in.

      Counter-posts, 12 in. × 12 in., and 12 in. × 14 in.

      Sill, 12 in. × 14 in.

      Intermediate caps and sills, 12 in. × 12 in., and 12 in. × 14 in.

      Sway-braces, 4 in. × 10 in.

Longitudinal braces, 6 in. × 8 in.

Purlins, 6 in. × 12 in.

Sub-sills, 12 in. round, flatted.

The trestle illustrated is built on a 10° curve. Mr. Joseph Hunter is the Chief Engineer of the road. For further description, see *Railroad Gazette*, February 6, 1891, p. 89. In the reduction of the drawing of this trestle the figures become so small that the reader is referred to the enlarged details for the dimensions which are also given above.

INDEX.

www.ingramcontent.com/pod-product-compliance
Lightning Source LLC
Chambersburg PA
CBHW021808190326
41518CB00007B/498